INDEPENDENT INNOVATION IN CHINA

Theory and Cases

Series on Innovation and Operations Management for Chinese Enterprises

Series Editor: Wu Xiaobo *(Zhejiang University, China)*

Editorial Committee: Xu Qingrui, Chen Jin, Wei Jiang, Gu Xinjian, Guo Bin, Huang Can

---

Vol. 1 Independent Innovation in China: Theory and Cases
by Zhong Huibo (Beijing Institute of Technology, China) and Lei Jiasu (Tsinghua University, China)

Series on Innovation and Operations Management for Chinese Enterprises – Vol. 1

INDEPENDENT INNOVATION IN CHINA

Theory and Cases

Zhong Huibo
Beijing Institute of Technology, China

Lei Jiasu
Tsinghua University, China

World Scientific

ZHEJIANG UNIVERSITY PRESS
浙江大学出版社

Published by

World Scientific Publishing Co. Pte. Ltd.
5 Toh Tuck Link, Singapore 596224
USA office: 27 Warren Street, Suite 401-402, Hackensack, NJ 07601
UK office: 57 Shelton Street, Covent Garden, London WC2H 9HE

and

Zhejiang University Press
No. 148, Tianmushan Road
Xixi Campus of Zhejiang University
Hangzhou 310028, China

Library of Congress Cataloging-in-Publication Data
Names: Huibo, Zhong, author. | Jiasu, Lei, author.
Title: Independent innovation in China : theory and cases / authored by Zhong Huibo and Lei Jiasu.
Description: New Jersey : World Scientific, [2017] |
Series: Innovation and operations management for chinese enterprises ; Vol. 1
Identifiers: LCCN 2016056694 | ISBN 9789813209893
Subjects: LCSH: Technological innovations--Economic aspects--China.
Classification: LCC HC79.T4 H84 2017 | DDC 338/.0640951--dc23
LC record available at https://lccn.loc.gov/2016056694

British Library Cataloguing-in-Publication Data
A catalogue record for this book is available from the British Library.

This edition is jointly published by World Scientific Publishing Co. Pte. Ltd. and Zhejiang University Press. This edition is distributed worldwide by World Scientific Publishing Co. Pte. Ltd., except China.

Typeset by China

Printed in China

Preface

(One)

Independent innovation as a very common phenomenon has existed throughout China's long history. The creation of the stone weapon during the stone age, followed by four great inventions and all of the technological innovations using modern technology and science during almost 40 years of the reform and opening up, all reflect the wisdom of the Chinese nation and show people the brilliant Chinese culture and its contribution to the world. No innovation, no progress of human society. No innovation from the Chinese nation, no advance of human civilization.

In the early 1990s, the concepts of "leading innovation" and "independent innovation" were put forward in China. However, not much attention was paid to them at that time in the Chinese society. Nowadays, it is independent innovation that calls much attention to China and not only represents a kind of confidence and determination in national development, but also has much deeper economic and political implications. The 17th CPC National Congress in 2007 clearly pointed out that "Enhancing the capability of independent innovation means constructing an innovative country"; and this Congress demanded that China insist on a policy of independent innovation with Chinese characteristics and that the country implement the concept of "enhancing the capability of independent innovation" in all aspects of the process of modernization. The primary goal in building an innovative country is the ability to significantly enhance independent innovation, so that the contribution of scientific and technological progress to economic growth can increase sharply and China will make it into the group of innovative countries by 2020.

Why didn't we emphasize the importance and urgency of independent innovation in the past as we do today? It is easy to find an answer by looking back on the history of innovation in modern and contemporary China to see that our nation has not had any achievements to match the four great inventions of the past for a long time. Today, people feel that inventions are as remote as ancient history. Are there any modern Chinese inventions that can shout to the world with confidence that they have truly been invented solely by us Chinese? No, not indeed! We do not remember ancient history; we merely remember the recent history of the Qing Dynasty. Due to the weak and ineffective Qing policies, the West attacked China without any strong counterattack from the Qing government. A few national industries were equipped with imported Western capitalist goods. The Westernization movement had the policy of "adopting Western culture, but giving it the backbone of Chinese culture." Despite bringing new ways of thinking and vitality to the Chinese feudal society that promoted the fundamental development of the national industry, it did not bring wealth and power due to the age-old feudal force. The Chinese nation was devastated in the Republican period between 1912 and 1949 with the unending civil and foreign wars. Though in the meanwhile, to some extent, our national capitalistic industry developed, few innovations appeared. Our national industry had made no significant progress. The majority of machine equipment was introduced from foreign countries.

In the first 30 years after the foundation of People's Republic of China, all kinds of social programs were created and began to develop in the right direction. Our country had built a more comprehensive economic system, especially concerning its industrial system, for example, electricity, the aeronautics and astronautics industry, the petrochemical industry, the auto and atomic energy industries had already been developed before 1978. Besides, the improvement of fixed assets and equipment also reflected the promotion of Chinese industrial technology. From the perspective of the industrial structure, it changed from mainly composed of primary industry to half primary, half secondary industry in about 30 years, from 1949 to 1978. The progress of industrialization changed the original situation of primary industry-dominant in China. The successful development of the atomic and hydrogen bombs, man-made satellites, modern aircraft, 10,000-ton hydraulic forging machines, and 10,000-tonners manifests the spirit of innovation of China.

After the Third Plenary Session of the 11th Central Committee of the Chinese Communist Party in 1978, we began to reform our economic system, and opened our domestic market to the world. Plenty of advanced foreign ways of thinking, technologies, and equipment were successively introduced. Thus, the whole national economy underwent a great transformation and was boosted. In the meanwhile, new problems appeared. Many foreign products have obtained a larger and larger share of our market since it opened up to foreign merchants. The market strategy and policy for technology do not bring us advanced international core technologies in every industry, therefore, China has become the processing base for industrialized countries and most profits flow into foreign pockets. For example, the price of our national DVD is less than 30 dollars, but two-thirds of the 30 dollars go to the patent fee of the foreign merchants. Our corporation gets only 1.5 dollars. We have a huge market with so many people, but our national corporations are not capable of taking the lead in the share of our domestic market. Lots of our national corporations are trapped in a vyious cycle, namely "introduction-imitation-incompetence-reintroduction." Therefore, we regret that we do not make exchanges with the advanced Western core technologies on our market. The year 2005 called on us to create a national strategy of independent innovation. All of the above are the reasons why the concept of "enhancing the capability of independent innovation, constructing an innovative country" was clearly put forward in the report of the 17th CPC National Congress in 2007.

(Two)

It deserves to be emphasized that there is no common view concerning the concept and theory of domestic independent innovation, though the words "independent innovation" have been written into relevant governmental documents for many years, and the whole society emphasizes and focuses on it persistently. Thus, the Chinese Ministry of Education has decided to support the Research Center for Technological Innovation of Tsinghua University to carry out theoretical research on Chinese independent innovation. This book is a main result of that theoretical research. This research has analyzed 9 cases of enterprises and 1 case of industry involved in independent innovations based on discussions of the fundamental problems of independent innovation, followed then by a summary.

This book has 6 chapters. The first chapter is an introduction, which explains the concept of "independent innovation" and the various ways it can be implemented. This chapter interprets some concepts related to independent innovation and introduces the outline of the entire book. The second chapter discusses the emergence of Chinese independent innovation, and then states that China must insist on the strategy of independent innovation, which is based on the practical enlightenment that we use our market to exchange for technology, that Japan established the fundamental practice through imitation, followed then by improvement, innovation, and development, and that the Republic of Korea has a kind of independent innovation based on introduction. Moreover, restraining factors of independent innovation are also pointed out in this chapter. The third chapter discusses the efficient path towards independent innovation based on the analyses of relevant cases. From these cases the path towards independent-innovation is concluded at the enterprise level from low-end to high-end (implementation, adoption, and re-innovation), the path towards independent innovation is concluded, based on the concept of knowledge-integration-innovation, and the path towards independent innovation goes from high-end to direct innovation (original innovation). The fourth chapter discusses the ability to construct, the potential mode, and the track to follow in order to reach independent innovation. This chapter involves the ability to innovate and the ability to innovate independently, the path towards enterprise innovation, and the way to achieve industry innovation, as well as how to achieve enterprise independent innovation and the ability to construct, the path towards industrial independent innovation, and the ability to construct. The fifth chapter discusses technology integration and the mechanism for enterprise independent innovation. Finally, this chapter discusses the relationship between technology integration and independent innovation, the content of and the procedure for achieving technology integration; it also talks about technology integration in two perspectives and introduces a feasible mechanism for enterprise technology integration. The sixth chapter discusses the institutional support for independent innovation, namely the construction of a system of innovation. This chapter involves the historical evolution of the Chinese system of innovation and the relevant institutional problems of constructing our country oriented towards independent innovation, as well as industry and enterprise innovation.

(Three)

The appendix to this book includes all enterprise cases and industry cases which are analyzed in the case-study research "the theoretical research of Chinese independent innovation." The enterprise cases are Huawei, CIMC, BYD, Tencent, Xunlei, Mindray, CHINT Group, Nuctech, Chery, etc. The industry case is the equipment manufacturing industry.

At the time of the publication of this book, we thank the support from Department of Social Science of the Ministry of Education and the Research Center for Technological Innovation of Tsinghua University,

Humanities and Social Sciences Base of the Ministry of Education.

Zhong Huibo
Professor at Beijing Institute of Technology
Lei Jiasu
Professor at Tsinghua University
Aug. 10, 2016

Contents

Appendix

Chapter 1

Introduction

1.1 The definition of the concepts related to independent innovation

1.1.1 The basic concept of independent innovation

Lei Jiasu et al. put forward the concept of "Lead Innovation" in 1993, and classified it as the Lead Innovation of internal technology and external technology. This concept actually implies the concept of independent innovation. Professor Chen Jin used the concept of "Independent Innovation" clearly in 1994. He thinks learning in the process of research and development is the dominant learning model of independent innovation, and the only way to grasp the essence of technology is research and development. In 1995, Professor Xie Xiezheng from Northeastern University (in China) proposed that independent innovation is technical innovation based on the transformation of scientific and technological achievements. In 1997, Yang Delin & Chen Chunbao pointed out that independent innovation means that enterprises rely on their own forces to do the activities of research, development, and technological innovation.

Independent innovation has 3 notable features: the first one is the independent breakthrough in core technology; the second one is the lead development in key technology; the last one is the lead development in the new market. In 1998, Professor Fu Jiaji pointed out that independent innovation is an innovative activity through which enterprises explore technological breakthrough singly or together, promote follow-up innovative links based on the above-mentioned technological breakthroughs, and then complete the commercialization of the technology and obtain commercial profit, so as to achieve the expected goal. Xu Guangyu (2005) holds similar opinions. In 2005, Zhou Jizhong et al. indicated that independent innovation is an innovative activity which means improving the ability of original innovation, integrated innovation and introduction, implementation, and adoption to obtain a number of independent intellectual properties and improve the competitiveness of a nation. In 2005, Liu Fengzhao et al. put forward that independent innovation is an innovative activity in which the subjects rely on their own (or rely mainly on their own) forces to achieve the breakthrough in science and technology (S&T), and then support and guide the development of economy and society and provide protection for the national security.

Independent innovation is a comprehensive concept. As to different subjects, independent innovation has different connotations. There are at least 3 aspects, namely country, industry, and enterprise. The independent innovation of a country mainly means the endogenous supply of key technologies which have common cross-industry characteristics, research and development of some cutting-edge technology as well as the lead in, or an influence on the fundamental direction in the field of basic science and applied science. The independent innovation of an industry mainly means the development and control of common-core technology that has a significant impact on industrial development, the participation of industrial organizations and/or enterprises in the establishment of international standards, and the right of industrial organizations to the independent pricing of products and services as well as the right to speak and have a dominant voice in value distribution on the market. The independent innovation of an enterprise means innovative activity that takes the independent intellectual property rights (IPR) and technology of an enterprise as the main symbol, and it takes integrated innovation and re-innovation on the basis of introduction as the main form of realization, in order to enhance the core competitiveness of an enterprise and shape an independent brand.

Wen Ruijun & Gong Jianli (2005) state that independent innovation can be divided into two categories according to the depth of innovation. The first is progressive independent innovation, namely establishing a new technological platform through the integration and introduction of a variety of existing technologies, so as to achieve innovation; the second is fundamental independent innovation, namely inventing a new technology through one's own research and thus developing a new product or a new generation of product. What they have in common is that innovators have technologies with an independent intellectual property and develop new products or technologies based on this.

Considering the above-mentioned knowledge and the reality of China, the book defines independent innovation as follows: Independent innovation is an innovation initiative in which innovators possess the core technology, independent intellectual property, and the participating right to establish international standards through independent research and development, imitation and innovation, integration and cooperation, and finally realize their business value.

1.1.2 The subjects of independent innovation

At present, the domestic debate on independent innovation focuses on the subjects of independent innovation. What/Who are the "subjects"? Considering its origin, the ancient Greek philosopher, Aristotle, used the term in the sense of logic. He thought that the subject of proposition, relative to a predicate, is a bearer of some characteristics, status, and effects. It indicates the existing entities in the world. The subject has consciousness, creativity, and initiative. In this sense, we can understand that an enterprise must be an "undertaker" and an "agent" of independent innovation. Therefore, independent innovation must emphasize the nature of a "subject," not the process of research and development. But the fact that the nature of an enterprise is complex is a big problem. Which enterprise can be regarded as a subject of independent innovation? It is still a question to be discussed. Domestic awareness of the question goes through the following stages.

In the first stage, the subject of independent innovation was equivalent to the subject of technological innovation. In 1994, when Professor Chen Jin from Zhejiang University used the concept of "Independent Innovation," he mainly emphasized the learning model of introducing technology to independent innovation. He thinks that learning in development is the dominant learning model of independent innovation. But he did not make it clear who the subject of independent innovation is. In 1995, Professor Xie Xiezheng from Northeastern University pointed out that with respect to the introduction of technology as "technological innovation of others," the "others" is the foreign technology. Professor Xie thinks that independent innovation means that independent innovation is "the technological innovation that is based on the transformation of scientific and technological achievements." He stresses the "ownership" of technology, but the subject of independent innovation is not clear. Subsequently, Dr. Yang Delin from the Graduate School of the Chinese Academy of Sciences, is the first one to have come up with a clear exposition of independent innovation. In 1997, he stated that independent innovation means that enterprises rely on their own forces and abilities to do the activities of research, development, and technological innovation. Its features are independent breakthroughs in core technology, leading the development of key technologies and being a leader in the development of the new market. It is easy to see that Dr. Yang Delin implies that the subject of independent innovation is the "enterprise."

In 1998, Professor Fu Jiaji also implied that the subject of independent innovation is "enterprises." Liu Fengzhao et al. (2005) viewed "independence" as a pattern of behavior distinct from the traditional way of introducing technology. They only mentioned "innovation subject," but did not make it clear who it is.

In fact, from a national perspective, the subject of independent innovation is the nation. While from the perspective of enterprises, the subject of independent innovation is the enterprise.

In particular, the discussion about the subject of innovation by Liu Wei, a professor of Peking University, should be mentioned. Liu Wei (2006) thinks that, under the system of planned economy, the subject of innovation is the country and its specialized agencies, because the main source of research funding comes from financial allocations. The main research teams are specialized scientific research institutes and universities who receive the financial support. But in the system of market economy, enterprises are the main market subject, so they should not only be the micro subject of resource allocation, but also the subject of technological innovation. From this point of view, the so-called independent innovation mainly refers to the independent innovation of enterprises, instead of the independent innovation of the nation. Of course, this does not mean that every enterprise needs to research and develop technology. Part of, even all of the required technology could be obtained through the market mechanism. But on the whole, innovation must be organized and executed by enterprises. It is not difficult to see that Professor Liu holds that the enterprise is the subject of technological innovation.

In the second stage, the subject of independent innovation was clarified further. In the previous stage, the scholars of the field agreed that the subject of independent innovation is the "enterprise." It is basically clear that the subject of independent innovation should be Chinese enterprises at the present stage. But a further question is how to define "Chinese enterprise." Is it confirmed by "capital controls," "place of incorporation," "possession of patent," or "ownership structure"? Obviously, it is not appropriate to identify a Chinese enterprise by "place of incorporation." At this stage, some Chinese companies are registered in foreign countries in order to obtain more commercial profit, but are controlled by Chinese capital. These enterprises certainly cannot be ruled out of the group of the Chinese subject of independent innovation. On the contrary, some companies are registered in China, but are not controlled by Chinese capital. These enterprises should not be regarded as the Chinese subject of independent innovation. Furthermore, making "possession of patent" the standard is not restrictive enough. Some Chinese companies apply for a patent in a foreign country in order to open its market abroad; an example is the "Laser Phototypesetting" invented by Wang Xuan which was first registered in France. So, identifying the Chinese subject of independent innovation by "possession of patent" is not appropriate.

In view of the discussion above, the authors of this book agree to take "capital control" and "ownership structure" as the criteria to identify whether an enterprise is the Chinese subject of independent innovation or not. Ownership (control) not only represents the decision-making power, but also represents usufruct. Some scholars have similar views. For example, Hu Jin (2006) believed that the Chinese subject of independent innovation refers to local enterprises under the control of Chinese capital. Similarly, *21st*

Century Business Herald gave out a "China software innovation list" on June 22, 2006. The candidate had to be a domestic enterprise that was a leader in industrial growth, having originality in a particular field and being full of significance in the field of Chinese software innovation. Visibly, only enterprises funded by local capital were believed to be subjects of independent innovation. Accordingly, as for joint ventures, enterprises controlled by foreign capital should not be regarded as the Chinese subject of independent innovation, since majority ownership is the right of an enterprise as a legal person to share property and interests. As for Sino-foreign joint ventures, it means majority shareholding when the foreign capital has more than 50% of the equity. In these enterprises, the Chinese shareholders don't have the decision-making power and the initiative of innovation. The innovation of enterprises is even strictly limited by the foreign shareholders. Shanghai Volkswagen is a typical example: it has been a joint venture for 20 years, and because the Chinese side does not have control power, it is very difficult for the enterprise to carry out independent innovation. Shanghai Volkswagen had attempted to improve its products by using the latest technology of the Volkswagen company, but the idea was soon rejected by Volkswagen, which has strict rules that Volkswagen technology cannot be used for any Chinese brand.

Since foreign holdings will make Chinese shareholders lose the right to independent innovation, it often leads to equity battles in Sino-foreign joint ventures. Taking Hangzhou Advanced Gearbox Group, formerly known as Hangzhou Gearbox Works, as an example. After the establishment of the Sino-foreign joint venture, an equity battle occurred immediately. The Chinese side proposed that the company should be controlled by themselves, but the foreign side insisted on their control of the company. Negotiations came to a deadlock, until the foreign side claimed that if the company were not controlled by them, they would not have invested in Hangzhou. Finally, Hangzhou Gearbox Works compromised under pressure and agreed that the foreign side should hold 70% of the shares. They signed a joint venture agreement in 2005. The agreement stipulated that the vehicle gearbox project and related products and technology equipment with market competitiveness should be combined in the joint venture company. Thus, Hangzhou Gearbox Works would not develop, produce, or sell marine gearboxes, engineering machinery gear boxes, truck gearboxes, and other related products. Neither could it use these trademarks. Thus, the most advantageous products and the strength of research and development of the factory were owned by the joint venture, and the weak projects were completely replaced by foreign products. Thus, the foreign side generally had a negative attitude toward independent innovation. Since the Chinese side lost control power, they could only be at the foreigners' service. The survival, innovation, and development of the company are controlled by others. In view of this, Chinese subjects of independent innovation can only be those enterprises that are registered in China under absolute control or with holding rights by local capital.

1.1.3 The dispute over the definition of independent innovation

In defining independent innovation, domestic debate mainly concerns the following aspects:

First, the dispute over the definition of "independent." Some scholars think, from a literal understanding of the word, that "independent" implies who advocates to innovate, how to achieve it, and who the result belongs to. The "who" of "who advocates" is the subject of independent innovation; "how to do about it" is the way to innovate, "who the result belongs to" is the resulting ownership of the innovation. That the enterprise is the subject of independent innovation is basically accepted, because the enterprise is the subject of investment, research and development, and the application of achievements. But some scholars believe that both enterprises and scientific research institutions are the subjects of independent innovation, and that enterprises occupy the leading position. Also some scholars hold that

"Independent Innovation" is relative to others. In the case of the entire interests of China, independent innovation means innovation activities of local enterprises, and the subject of independent innovation should be the local enterprises controlled by local capital.

Second, the dispute over the contents of independent innovation. Chen Jin (1994) believed that independent innovation is "independent technological innovation," which is a specific developmental stage of an enterprise in the progress of the introduction, implementation, and improvement of foreign technology. Yang Delin & Chen Chunbao (1997) believe that when independent innovation is used for enterprise technological innovation, it refers to the fact that enterprises rely on their own forces to carry out independent research and development, and carry out activities of technological innovation. But some scholars oppose this idea claiming that independent innovation is limited to technological innovation. For example, Xie Xiezheng (1995) does not agree that independent innovation is "written" as independent technological innovation. He holds that independent innovation includes innovation of organizational management, the choice of invention, and technological achievements that are gradually suitable for production applications, so as to obtain success on the market, and it also includes the efforts of developing technology according to the market and production needs. Wang Yuan (2005), President of the Chinese Academy of Science and Technology for Development, starts out from the relationship between independent innovation and the market. He sees independent innovation as a process of economic development. He thinks that it is not only a pure technological invention, but also a process of economic and market realization. The essential characteristics of independent innovation are transformation of scientific technological achievements and the process of industrialization.

Third, the dispute over the results of independent innovation. The dispute over the results of independent innovation concentrates mainly on whether independent innovation can form independent intellectual property. Intellectual property is a stage in the achievement of innovation. It is created through creative human activity, and then it can be put onto the market through production and sale. In enterprise innovation, intellectual property plays a very important role. If there is no intellectual property, the enterprise can only function at the bottom of the world's "food chain." Real core technologies cannot be bought; only the strong ability of scientific and technological innovation can result in the formation of independent intellectual property and improve the international competitiveness of our country. Based on these considerations, independent innovation means owning core technology and intellectual property, and even breaking the technological monopoly of developed countries. However, some scholars think, from another point of view, that independent innovation should not be equated with independent intellectual property. Imitative innovation is namely "second innovation." Though the breakthrough of core technology relies on our own forces and abilities, it does not mean independent intellectual property. For example, technological improvement in Japan was achieved through imitative innovation, which also produced independent intellectual property.

Fourth, the disputes over the criteria used to identify the success of independent innovation. As Tsinghua University Professor Fu Jiaji pointed out in 1998, independent innovation is an innovative activity that enterprises explore within technological breakthroughs on their own or together through research, and then promote a follow-up link to complete the commercialization of technology, so as to obtain commercial profit. He believes that the criterion for the success of independent innovation is the creation of business value. Similarly, the British scholar Freeman (1987) holds that the failure of innovation is the failure to establish an effective market or to gain profit. Accordingly, successful innovation means that an enterprise obtains significant market penetration or profit. Thus, Freeman (1987) does not regard commercial profit as the only sign of successful innovation. At the same time, he thinks that the highest level of independent innovation is achieving independence of the local enterprises' capability for innovation. The enterprise and the nation have continuous learning ability and organizational capability.

Fifth, the dispute over the key factors regarding the success or failure of independent innovation. What are the key factors for the success of independent innovation? Some scholars think that innovation requires courage and wisdom. Talent is the key to independent innovation. Also some scholars hold that a strong innovation awareness and cultural depth form the foundation for innovation. Other scholars believe that independent innovation is the endogenous variables of learning abilities and organizational capabilities which are closely linked to multilevel economic opening up. Basic research is the source of independent innovation.

1.2 Classification of the models and types of innovation

Innovation has a great variety of characteristics and uncertainties, but it also has some established rules. The successful model of innovation is a kind of rule, thus, the academic circle and the business circle never stop researching and discussing the models of innovation.

1.2.1 Foreign classification of innovation models based on different paths

(1) "The fifth-generation model based on the path driven by innovation" proposed by Rothwell (1994)

This is the most representative explanation for the evolution of the innovation model in the foreign academia. Rothwell (1994) thinks that there are five stages that the evolutionary path of innovation has been through: i) Industrial technological innovation in the 1950s. It is mainly the linear progression of "technology push," namely the "technology push model." ii) Technological innovation from the late 1960s to the early 1970s. It is mainly the "market pull model," emphasizing innovation according to the market demand. iii) In the late 1970s, the two models began to integrate, which led to the emergence of the "interactive model of innovation between technology push and market pull." iv) From the late 1970s to the early 1980s, there emerged the innovation model of joint participation by the departments of technology, design, manufacture, management and marketing, namely the "integrated innovation model." v) Technological innovation in the late 1980s, when the "systems integration and networking model" appeared. This innovation model was a parallel process of system functions, such as research and development, prototype production, manufacturing, etc. It especially emphasized the innovation networking of close cooperation among companies.

(2) "The classification of innovation models based on the evolutionary path of market competition" proposed by Kumpe & Bolwijn (1994)

They think, from the perspective of the market competition path, that innovation can be divided into four models based on: i) Efficiency competition. This is the price competition model of the closed seller's market until the early 1960s. ii) Quality competition. In the 1970s, the competition among firms was basically quality competition and price competition. The market pull led to a more integrated view of products and processes, and more attention was paid to producibility and serviceability, and closer cooperation between sales, marketing, research, and development. iii) Flexible competition. The 1980s was an era of flexible firms. The organizational design of flexible firms was based on simplification and the creation of fast feedback loops and innovation efficiency. Manufacturability became all-important. The flexible firm focused on strategic technological innovation. Most "open" R&D activities were replaced by contract R&D. iv) Innovation competition. In the 1990s, the most successful firms were those that were innovative. They always struck the right balance between renewal and stability.

(3) Some scholars from newly industrialized Asian countries categorize the innovation model from different angles

Represented by the Korean scholars Lee & Lim (2001), they divide the innovation model into three types: Path-following catching-up, path-skipping catching-up, and path-creating catching-up.

1.2.2 Domestic classification of innovation models seen from different angles

According to different innovation objects, the famous scholar Fu Jiaji (1998) divides innovation into product innovation and process innovation; according to the intensity of technological change in the innovation process, he divides innovation into radical innovation and incremental innovation; according to the source of the enterprise's R&D resources and ability, he divides innovation into independent innovation, imitative innovation, and cooperative innovation. The classification method and discussion of the advantages and disadvantages of independent innovation have been cited by many colleagues and scholars. Some scholars have developed further research on the innovation model based on this representative method, as follows.

(1) The classification based on innovation object, such as technique, product, culture, etc.

Li Gang & Chen Changbai (2006) divide innovation into five categories: Innovation of basic invention; innovation of core technology; innovation of improvement; innovation of product design; innovation of culture.

(2) The classification according to the intensity of technical change in the innovation process

Zhang Hongshi & Chen Jin (2005) divide innovation into breakthrough innovation and incremental innovation. Wang Shenghui (2007) divides innovation into non-breakthrough innovation and breakthrough innovation. According to the different emphases on product component innovation and the different degrees of product structure innovation, he divides non-breakthrough innovation into product innovation of adaptation, integration, and upgradation.

(3) The classification according to technology sources

i) Yang Shuiyang (2005) divides innovation into three types, namely original innovation, imitative innovation and integrated innovation; ii) Wang Yusuo & Luo Yongtai (2007) also list three categories in the study of high-tech industry innovation: imitative innovation, original innovation and integrated innovation of the alliance of university and enterprise; iii) Lyu Yuhui (2007) divides innovation into independent innovation, imitative innovation, introduced innovation and cooperative innovation.

(4) The classification based on empirical research

By summarizing the innovation experience of TCL and other famous enterprises based on the technology roadmap of introduction→implementation and adoption→improvement→independent innovation, Wang Miao et al. (2005) put forward the following innovation model: i) The innovation model of improvement and promotion around production technology and process; ii) the innovation model around the source of elements; iii) the innovation model around the market; iv) the innovation model around high-tech inventions. According to the different ways of obtaining enterprise technology, Luo Jiujiu et al. (2007) divide innovation into original innovation and secondary innovation; according to the different participants in innovation, they divide it into cooperative innovation and independent innovation.

It is not difficult to see that the above-mentioned types and models of innovation are classified based on various angles of research, so they are often confusing and even overlapping.

1.3 Classification of independent innovation

At present, there are many classification methods regarding independent innovation, which can be divided into four kinds: i) Based on the technological characteristics of innovation, independent innovation is divided into original innovation, integrated innovation, and re-innovation based on introduction, implementation, and adoption; ii) based on the characteristics of innovation activity, it is divided into conventional innovation and unconventional innovation; iii) based on the perspective of the technological breakthrough, it is divided into dominant breakthrough design innovation, core components innovation, and product architecture innovation; iv) open innovation, especially cooperative innovation, is often regarded as the important organization model of independent innovation.

1.3.1 Original innovation, integrated innovation, and re-innovation based on introduction, implementation, and adoption

The concept of independent innovation occurred domestically first in the 1990s. At that time, "independent innovation" was used as the opposite of technology introduction. Therefore, independent innovation was regarded as innovation based on one's own technology first; and, re-innovation based on the introduction, implementation, and adoption of the foreign technology was regarded as independent innovation as well. At the same time, innovation was divided into independent innovation and imitative innovation. While some scholars equated independent development with independent innovation, and regarded independent development of products with independent intellectual property as independent innovation. In recent years, people have gradually realized that the core of independent innovation is the formation of independent intellectual property. And further, both independent development and re-innovation based on introduction should be regarded as independent innovation, as long as they have formed all or part of the achievement of intellectual property. For example, in February 2005, the then Vice Minister of the Ministry of Science and Technology, Dr. Shang Yong, proposed that independent innovation should focus on: original innovation, integrated innovation, and re-innovation based on the introduction, implementation, and adoption of the foreign technology. In July 2005, during the plenary session of the leading national S&T group meeting, then Premier Wen Jiabao stressed that more attention should be paid to emphasizing and pushing independent innovation forward and to the improvement of our ability to achieve original innovation, integrated innovation, and re-innovation based on introduction, implementation, and adoption. The types of independent innovation were initially confirmed. Since then, people have agreed on a preliminary classification of independent innovation based on technological characteristics.

(1) Original innovation

When Professor Walz (1995) of the University of Mannheim in Germany analyzes economic growth, he emphasizes endogenous innovation and imitative innovation. Considering the basic meaning, he says that endogenous innovation is original innovation. He believes that endogenous innovation means that the R&D of key technology is carried out completely by the enterprises themselves, which forms independent intellectual property; endogenous innovation is the basic embodiment of an enterprise's independent innovation ability; the capability for endogenous innovation is the basis for the enterprise's cooperative innovation or re-innovation based on implementation and adoption. In other words, other innovative subjects tend to cooperate with the enterprise which creates technology accumulation and ability promotion through endogenous independent innovation. Thus, the enterprise will have the ability to complete re-innovation based on implementation and adoption of the technology that has been introduced.

(2) Integrated innovation

Integrated innovations, namely new products, are developed based on new technology which is formed through the organic integration of existing technology around specific market demand. A typical example is the photocopier. Before the photocopier appeared, all of the technologies it uses had already emerged, and they were very mature. The innovators of the photocopier just integrated all of the existing technologies, which formed the systematical technology of the copiers. Moreover, Shanghai Zhenhua Port Machinery applied the US GPS system to the container crane in 2000 and solved the problems of crane flaking and container management. A year later, the group applied for a patent on a global scale, and became the industrial leader in technology development. In addition, the Chinese Shenzhou manned spacecraft project is also a model of integrated innovation.

(3) Re-innovation based on introduction, implementation, and adoption

Re-innovation based on introduction, implementation, and adoption means first learning to use mature foreign technology after its introduction, followed by implementation and adoption, then improving and innovating it to some extent. Examples of this have been very common in China since the reform and opening up, especially in the fields of household electrical appliances and military equipment. For example, Haier and other household electrical appliance enterprises introduced foreign refrigerator technology and used it effectively, soon achieving significant improvement and innovation. Then Haier products have been selling well both domestically and abroad. Another example is in the military industry: China introduced new-generation warplane technology from Russia, first manufacturing, then improving. Then China produced the same type of fighter plane which is far better than the Russian fighter, and it has won over a part of the international market.

1.3.2 Conventional and unconventional independent innovation

As mentioned above, based on the characteristics of innovation activities, independent innovation can be divided into conventional and unconventional innovation. Conventional innovation occurs mainly in large and mature enterprises, while unconventional one mainly occurs in small entrepreneurial enterprises.

(1) Regular independent innovation of large and mature enterprises

Many large and mature enterprises have accumulated rich experience in independent innovation. These enterprises often regard independent innovation as normal conscious activities. Many large Chinese enterprises have formed the new paradigm of conventional independent innovation that has its own characteristics, such as Haier, Hisense, Huawei, and others. The conventional independent innovation of large enterprises has the following characteristics: The first is internalization of technology sources. They build a central research institute inside the enterprise in order to make the R&D process internal, manageable and controllable. The second is the functionalization of creative activity. The entrepreneur's contribution to innovation diminishes in innovation activities which are completed by a special trained expert group or departments. The third is the reutilization of the department of resource input for enterprise planning which decides the R&D funding. The proportion of R&D investment relative to the operating income is relatively stable, so as to ensure that the enterprise has sufficient R&D investment. The fourth is the predictability of the results of innovation. Based on technical accumulation and experience, large enterprises can often make an objective prediction for an independent innovation. The fifth is the succession of innovations. Product innovation in large enterprises has a strong correlation with their old products. So does the process innovation. Another characteristic is the emphasis on continuous improvement of technology and products that large enterprises focus on, such as increasing product usability to the users, improving the process reliability, etc.

(2) Unconventional independent innovation in small entrepreneurial enterprises

In reality, unconventional independent innovation is mostly driven and carried out by small entrepreneurial enterprises. This kind of independent innovation has the following characteristics: The first is the close contact between innovation and entrepreneurship. Once the new technological invention proves to be repeatable, it is possible for the inventor to create a new company to make the corresponding technological invention become new products that the market demands. The second is that the innovation often comes from the entrepreneur. Hence the innovator is often the entrepreneur. Some examples are: Deng Zhonghan, founder of Vimicro; Li Yanhong, creator of a Chinese search engine; Huang Ming, CEO of Himin Solar, who created a solar city; Qiu Hongyun, a participant in the creation of Seemile Company. They are "entrepreneurial entrepreneurs." The third is that many innovative achievements exert a revolutionary influence or create a new industry or a new market. This kind of innovation is often a breakthrough, and even makes a disruptive change in the market structure.

1.3.3 Breakthrough innovation of dominating design, core components, and product architecture

As has been noted, based on the perspective of technological breakthroughs, independent innovation can be divided into innovations of dominating design, core components, and product architecture, respectively.

(1) Innovation of dominating design

Technological development is not continuous. The dominant designs of similar products in different periods may be different. According to the technological aspect, the dominant design can often be expressed as the product design scheme and normal innovation that keep developing along an established technology trajectory. The British economist Dosi (1982), who lives in Italy now, has proposed the theory of a technology trajectory. He thinks that a technology trajectory is a regular problem-solving process within a specific technological paradigm. A technological paradigm is a general outlook on the technological problems by specific scientific principles and materials technology. For example, when transistors replaced the electron tube in the era of the integrated circuit, the increasing degree of integrated circuit became a particular trajectory of technological development. Because the change of the dominant design is often the application of a new scientific principle, it needs a system supported by both basic and applied research. It may form the basis of a patent and/or a key patent, which is the protection for the new-generation dominant design. Regarding the market, the dominant design is the consensus of the expectations that the manufacturers and the customers attribute to the product. For example, although the Dvorak keyboard, invented in 1936, is faster than the QWERTY keyboard, invented in 1868, by almost 5%, the QWERTY keyboard has dominated the market in competition due to the earlier acceptance in the market. Based on this, the learning effect and the dominant design will show a "trend of increasing returns," and may affect the trend of the next generation with the "lock-in effect."

(2) Innovation of core components

The core components (i.e. the core parts in mechanical products, the same below) is a key part of the whole product, an important technical support for the realization of specific functions of the whole product and the key to the realization of the dominant design. From the perspective of system theory, the whole machine is a big system or even a large complex system, the components are the secondary system, and the core components are the core of the secondary system. The core components reflect the core idea of the whole product design. An example is the automobile, where the gasoline engine obtains kinetic energy by burning in the engine so as to realize the core design of automobile driving. It has a different principle from that of the power system of the pure electric vehicle, the fuel cell vehicle, and the hybrid vehicle. Core component technology is complicated, but it can form a technological platform that provides a series of support systems for the end product, such as cameras, copiers, laser printers, and image scanners of the Japanese company, Canon. The core element of these end products is the "optical engine," which is based on precision optics, imaging technology, the microprocessor, and precision machinery. Thus, if there is an innovation in the core component, it can provide technical support for the improvement of the present dominant design and the new dominant design. For example, the central processor is mainly responsible for the operation and control of a computer, directly determining the speed of computing. Therefore, only from 1993 to 2000, it experienced four generations of products, that is, Pentium®, Pentium® II, Pentium® III, and Pentium® 4. With its integration level increasing from 3,100,000 to 42,000,000, the speed of computing was greatly accelerated. Of course, the supporting effect of the core component also depends on the matching degree between the core component itself and the product architecture as well as the coordination degree between the core component itself and the other components.

(3) Innovation of product architecture

Product architecture is the way the product components or parts are connected. It decides the function and the performance of the machine product by the system's mechanism. In different product architectures, the product systems will show different functions or performances even with the same components or parts. This is similar to the way vehicle architecture expresses the connecting relationship between each function module, which decides the vehicle load, controllability, comfort, and other functions, and how it determines the reliability, economy, and other performances of the product. Product architecture technology is the way the components and parts are combined and connected. It is a knowledge system for effective functioning of the overall system. Through product architecture innovation, innovators can turn product concepts into specific components of the whole product combination, which may form a more valuable product. And if innovators want to master architecture technology, they must understand the characteristics of the various components, parts, and interrelated mechanisms. Some architecture technology will even change the whole system's function or performance. Architecture technology in product innovation may provide better probability. On the one hand, product architecture has the potential to optimize the overall performance of the product. Without any change in components or parts, gradual product architecture innovation may improve the overall performance of the vehicle. On the other hand, innovation of end products is often the result of technological architecture innovation. In companies such as Sony, in Walkman product innovation, the core components (micro motor, slice rechargeable battery) are invariable. They only changed the architecture products locally and the non-core components, so as to optimize some auxiliary functions and performances. Only from the 1980s to the early 1990s did Sony launch 572 products on the American market: Products such as those making the Walkman more compact in order to meet the demands of children, or lighter with an anti-vibration function to meet the demands of sports, or even improved in its waterproof performance so that people could use it on the beach.

1.3.4 Independent innovation based on cooperation

As stated earlier, based on the organization characteristics of innovation, business and academic circles often regard open innovation, especially cooperative innovation, as an important organizational model of independent innovation.

In 2003, through the study of Qualcomm, Genzyme, Chicago Tribune, Procter & Gamble, and other enterprises, Henry Chesbrough, professor at the Harvard Business School, found that these enterprises would never depend entirely on internal R&D. They had turned to a model that was more open to obtaining technology. He put forward the concept of "open innovation" based on his study, which is "valuable ideas" can be simultaneously obtained from both outside and inside the company, and its pathway towards commercialization can be carried out both inside and outside the company. The open innovation model is becoming an effective model to quicken innovation for some innovative enterprises. Among them, cooperative innovation is the most specific and an important open innovation model. The so-called cooperative innovation is among enterprises, scientific research institutes and universities. It is based on the common interests of partners with the promise of sharing resources or complementary advantages. There is a clear cooperative target, time limit and rules; all parties of the cooperation in the whole process or in some links of it invest and participate in the same way, and share the results and risks. In reality, the key to cooperative innovation is handling fruit, ownership, and common property well. In China, cooperation is the most typical model of independent innovation. To a considerable extent, in recent years, China has achieved many valuable independent innovations based on the enterprise-university-research cooperation.

1.4 The basic layout of this book

The foundation of this book is empirical analysis, with the purpose to construct a theory of independent innovation based on our national conditions and to clarify the concept and connotation of independent innovation based on theoretical research on typical cases.

This book is divided into six chapters: The first chapter is the introduction, where the concept of independent innovation is defined and clarified, and the main types of independent innovation are set forth. The second chapter discusses the urgency for implementation of the Chinese strategy of independent innovation. Its practical implications are: China exchanges market for technology; Japan laid the foundation for imitation and the development of innovation; the Republic of Korea laid the foundation through introduction and carried on into the future with innovation. Then, this book affirms that China must adhere to the national strategy of independent innovation and it points out the relevant constraints on China's independent innovation. The third chapter discusses the effective path of independent innovation in China. Based on the case analysis, there is a summary of the paths towards independent innovation where enterprises carry out innovation activities from low-end to high-end (implementation, adoption, and re-innovation) in China, or through integrated innovation based on knowledge (integrated innovation), and innovation directly from the high-end (original innovation).

The fourth chapter discusses the building of independent innovation capacity, the model and path of Chinese independent innovation, which involves the ability for innovation and independent innovation, the model of enterprise innovation and the path towards industrial innovation, the construction of a model and the innovation ability of enterprises, the construction of a path and industrial capability for independent innovation. The fifth chapter discusses the technology and mechanism of the method of integration in the independent innovation of enterprises, involving the relationship between technological integration and independent innovation, the content and process of technological integration, technological integration from two angles, and a feasible mechanism of technological enterprise integration. The sixth chapter discusses the guarantee system of Chinese independent innovation, namely the construction of a system of innovation, involving the historical evolution of China's national system of innovation and the problems related to the construction of a system of independent innovation oriented to the system of country, industry and enterprise innovation.

References

Chen, J. The learning mode from technology introduction to independent innovation. *Science Research Management*, 1994, 15 (2): 11-16.

Dosi, G. Technological paradigms and technological trajectories: A suggested interpretation of the determinants and directions of technical change. *Research Policy*, 1982, 11 (3): 147-162.

Freeman, C. *Technology Policy and Economic Performance.* London: Pinter, 1987.

Fu, J. *Theory on Technological Innovation.* Beijing: Tsinghua University Press, 1998.

Hu, J. Enterprise independent innovation research. *Theory Monthly*, 2006 (11): 155-157.

Kumpe, T. & Bolwijn, P.T. Toward the innovative firm: Challenge for R&D management. *Research-Technology Management*, 1994, 37: 38-44.

Lee, K. & Lim, C. Technological regimes, catching-up and leapfrogging: Findings from the Korean industries. *Research Policy*, 2001, 30 (3): 459-483.

Lei, J., Du, Y. & Deng, X. The difficulties of industrial growth in western China. *China Industrial Economics*, 1993 (8): 41-46.

Li, G. & Chen, C. The selection of enterprise independent innovation mode. *Science & Technology and Economy*, 2006 (1): 26-28.

Liu, F., Pan, X. & Shi, D. Research on the evaluation of regional independent innovation ability based on set pair analysis. *China Soft Science*, 2005 (11): 83-91.

Liu, W. The subject of independent innovation is large enterprises. *New Economics*, 2006 (1): 118-120.

Luo, J., Tian, X. & Kang, L. Research on and selection of enterprise independent innovation model. *Science & Technology Progress and Policy*, 2007 (5): 25-31.

Lyu, Y. & Ding, C. Technological innovation ecology and ecological technology innovation. *Science and Technology Management Research*, 2007 (3): 33-35.

Rothwell, R. Towards the fifth-generation innovation process. *International Marketing Review*, 1994, 11(1): 7-31.

Walz, U. Endogenous innovation and imitation in a model of equilibrium growth. *European Journal of Political Economy*, 1995, 11

(4): 709-723.
Wang, M., Hu, B. & Jiang, Z. Enterprise independent innovation pattern and strategy in the process of new industrialization in our country. *Economic Review*, 2005 (10): 73-78.
Wang, S. On after-breakthrough independent innovation patterns of enterprises. *Scientific Management Research*, 2007, 25 (3): 34-37.
Wang, Y. & Luo, Y. Research on high and new technology enterprise independent innovation model. *On Economic Problems*, 2007 (4): 63-65.
Wen, R. & Gong, J. Research of enterprise independent innovation ability evaluation. *Group Economy*, 2005 (7): 28-33.
Xie, X. The progress of science and technology, independent innovation, and economic growth. *Software Engineer*, 1995 (5): 33-38.
Xu, G. Research on high technology enterprises for independent innovation based on the technological trajectory. *Science of Science and Management of S.&T.*, 2005 (3): 148-152.
Yang, D. & Chen, C. Imitative innovation, independent innovation, and high-tech enterprise's growth. *China Soft Science Magazine*, 1997 (8): 107-112.
Yang, S. The theory foundation and basic patterns of independent innovation. *Industrial Technology & Economy*, 2005 (7): 66-70.
Zhang, H. & Chen, J. Research on the organization mode of the radical innovation. *Studies in Science of Science*, 2005 (4): 566-571.
Zhou, J., Zhang, L. & Tang, C. About the linkage between the independent innovation and intellectual property rights. *Management Review*, 2005 (11): 56-62.

Chapter 2

The urgency for implementation of independent innovation strategy in China

2.1 The practical enlightenment of "market for technology" in China

Since the reform and opening up, the Chinese government has introduced the important strategy and policy of "market for technology" to change the situation of backward technology in most industries. The so-called "market for technology" means obtaining advanced foreign technology for China's national enterprises at the cost of allowing foreign investors to sell large quantities of their products, to win the project bidding, and to establish sole proprietorship or partnership in China. China has been implementing the strategy and policy of "market for technology" for 20 years. Now foreign investors are involved in many industrial fields and their market share shows a growing trend. Therefore, it is not difficult to understand the urgency for enforcement of the independent innovation strategy in China by rethinking the positive and negative effects of "market for technology" in the progress and development of industrial technology.

2.1.1 Positive effects of "market for technology" on the innovation of Chinese domestic enterprises

By implementing the strategy and policy of "market for technology," the innovation of domestic enterprises has the following positive effects.

(1) Promoting technological development and improving the management level of many industries in China.

In industries such as the automobile, the generator, program-controlled switches, the mobile phone and the computer, "market for technology" can provide the industries with technological learning and accumulation of experience, improve the overall technical capability of these industries, and promote their development. Due to the force of political pressure and market interest, the technology for investment of foreign enterprises is often more advanced than the technology for sale. So some domestic enterprises have managed to master the advanced technology at a lower cost and in less time by implementing the "market for technology" strategy. Some technologies even fill the national gap and narrow the gap between domestic and foreign technology, which has laid the foundation for the development of related industrial technology in China. For instance, cooperating with the world's advanced automotive manufacturers has helped Chinese auto manufacturing enterprises to gradually master the technological methods and the process of vehicle design and production; moreover, it has also helped Chinese auto enterprises to accumulate valuable experience to exploit for independent innovation. "Market for technology" has promoted the development of many industries in China and it has also improved the management level of Chinese enterprises. One of the most important advantages of the cooperation between local capital enterprises and foreign enterprises is the advanced foreign management experience we have learned. It plays a very important role in improving Chinese enterprises' production, management abilities, and management level.

(2) Promoting the accumulation of industrial technical experience and the training for a skilled workforce

For example, the computer, the mobile phone, the auto and other domestic industries lacked necessary experience in technology and production. Accumulation of experience is very important. In the process of "market for technology," in order to make Chinese domestic enterprises or joint ventures produce qualified products, foreigners provide multi-layered training for Chinese staff, so as to guarantee that they have at least the capacity to use the equipment and produce normally. If there had not been an accumulation of cooperative study through the telecommunications research institute for more than ten years, it would not have been possible for Datang Corporation to rapidly develop the digital program-controlled switches after the American enterprise ITTI had provided their integrated circuit technology; without the accumulation of experience in producing computers, Lenovo and other PC companies would not have been possible to master computer technology quickly. Thus, the accumulation of experience regarding industrial technology is a necessary condition for industries to develop and for technology to progress. At the same time, due to the cooperation between domestic enterprises and foreign capital within the "market for technology" policy, a large number of experienced employees were trained, who systematically received the technical training or worked together with foreign experts, acquiring valuable experience from them. In short, due to cooperation, many Chinese technicians get practice and enhance their skills and therefore are able to replace or partially replace foreign experts in carrying out the corresponding work.

(3) Promoting the development of domestic industry and guiding the innovation of domestic enterprises

The implementation of the "market for technology" strategy has reduced some barriers for some foreign industries (such as the auto industry, the electric generator industry, and others) for entry into China, which gives Chinese national enterprises the opportunity to understand the world, the management experience and the advanced technology of developed countries. The entry of foreign capital has intensified industrial competition, enhanced the sense of crisis of Chinese enterprises, and stimulated them to improve their concept of management and their level of technology in order to perfect their products and services. It has also broken the monopoly of foreign products in some product fields, which has thus made it possible for foreign companies to export their technology to China. This plays an important role in the establishment and development of Chinese local enterprises. At the same time, joint ventures and cooperation with foreign investors have also helped some domestic industries to extend their industrial chain. Especially the increase in production and localization rate provides a broad market for the downstream enterprises of some industrial chains. According to experts' estimation, direct degree of association between the value of the auto industry and that of other related industries is 1∶2, and the indirect degree of association reaches 1∶5. The production value of the auto industry can bring a growth of 2.5 times the actual value of related industries. If the auto industry increases by 1 yuan, it will bring the upstream industry an appreciation of 0.65 yuan, and the downstream industry 2.63 yuan. All these aspects give a certain guidance to the innovation of Chinese domestic enterprises.

2.1.2 Negative effects of "market for technology" on the innovation of Chinese domestic enterprises

The implementation of the "market for technology" strategy and policy has the following negative effects on the innovation of Chinese domestic enterprises.

(1) Making some local enterprises strongly dependent on foreign technology

Although the policy of "market for technology" has brought some advanced technologies to China, it has made some local capital enterprises dependent on foreign technology. Especially, there has only been some marginal technology transfer of foreign enterprises to some industries. We did not obtain or master core technology. What we can only do is to buy foreign products or technologies once in a while. After the establishment of joint ventures, some domestic enterprises even closed the original research institutes and abandoned the original research efforts. At the same time, some leaders overrated foreign technology, resulting in the tendency of some domestic owners to choose foreign technology and equipment in project bidding. This further increases domestic enterprises' dependence on foreign technology.

(2) Not conducive to cultivating independent brands

Under the format of "market for technology," China has established numerous joint ventures. These enterprises often import foreign technology, equipment, and even production lines. They use cheap Chinese labor and natural resources to produce. Many joint ventures, such as Shanghai Bell, Shanghai Volkswagen, etc., use their foreign partner's brand rather than establishing an independent brand because of their own weak brand recognition and contract limitations. Obviously, this is not conducive to cultivating the brands of local enterprises.

(3) Increase in enterprise operating costs and decrease in innovation investment

Since China implemented the "market for technology" policy, originally many local enterprises hoped to "exchange the market for technology." Some enterprises gave up the market and worked with foreign capital, but they did not get the advanced and applicable technology they had hoped for. As a result, these enterprises had to turn back to their own development of required technologies, which not only increased their operating costs, but also weakened their ability for innovation investment. There are also some enterprises, which have obtained the foreign advanced and applicable technologies but made no effort to implement and adopt foreign technology. Thus, they have fallen into the vicious cycle of "introduction–falling behind–reintroduction–falling behind again." On this background, in order to keep their technology on par with their competitors, many enterprises have had to pay for the introduction of technology or equipment from abroad every few years, which further weakens their ability in innovation.

2.1.3 The effect of "market for technology" requires independent innovation

Historically, the policy of "market for technology" was in accordance with the general law of market economy; "market for technology" has successful examples in foreign countries. Japan and the Republic of Korea had practical experiences of "market for technology" after World War II (see the following sections). In particular, the "market for technology" strategy went along with China's condition of seeking development after the implementation of the reform and opening up from the late 1970s to the early 1980s. At that time, for 10 to 20 years, a lot of industries in China fell behind European countries, America, Japan, etc. The gap was even greater in some industries. After the opening up, competing with their foreign counterparts, the enterprises had to make every attempt to bridge the technology gap. After they failed in purchasing foreign technology directly, the policy of "market for technology" became the inevitable choice at the beginning of the 1990s. Of course, the implementation of "market for technology" had different results in different industries. The reasons are the difference in policy and supervision of the relevant governmental departments and the difference in the implementation of the policy in different enterprises. Some deviations also appeared in the implementation of this strategy and policy. However, at the present stage, China needs to develop through independent innovation.

(1) At the beginning of the 21st century, the Chinese central government duly proposed the strategy of independent innovation

In the process of "market for technology," some developed countries are hindering our development of technology. At the same time, excessive dependence on the technology of developed countries also intensified the contradiction between demand and supply of technology in China. Although the globalization of the economy and the internationalization of S&T are increasingly deepening, the international flow of S&T resources is becoming increasingly commonplace; the basic attitude of developed countries towards the implementation of a technical lock-in against China will be hard to change for a long time. "Batumi" and later the "Wassenaar Arrangement" are the cases at hand. The channel for China to obtain advanced technology from the outside is getting narrower and narrower. In this context, China needs to adjust the "market for technology" strategy, which means continuing to obtain foreign technology and making a greater effort to meet the need for technology in the process of economic development through independent innovation. Based on this, the independent innovation strategy proposed by the Chinese central government at the beginning of the 21st century is providential. It is also in agreement with the fundamental realities of the country after over 30 years of reform and opening up[1].

(2) Independent innovation does not exclude the continued use of advanced foreign technology in view of the benefits it can bring us

The implementation of the independent innovation strategy in China is not aiming at rejecting the advanced foreign technology which can benefit us. To be exact, what we should do is to seek a more favorable balance between independent innovation and the use of advanced foreign technology. In some fields which are related to national security, the foundation for economic and social development, and the international industrial competitiveness and core technology, we must insist on independent innovation; in the field of non-common and non-core technology, we should actively make use of international technological resources. For the resource allocation of independent

[1] It was less than 30 years after the reform and opening up when the Chinese central government put forward the strategy of independent innovation.

innovation and the use of international technology, on the one hand, we should allocate more resources to independent innovation; on the other hand, in the process of using international technological resources, we should also give more resources to the link of implementation and adoption rather than that of introduction. Only when the technology has been truly adopted and transformed to the technological ability of our local enterprises and institutions, can we talk about re-innovation and connecting the international technical resources and independent innovation effectively, so as to promote China's sustainable development for a long time.

2.2 Japan: Imitation, improvement, and innovation

Japan and the Republic of Korea are both countries that have achieved an economic take-off after World War II. The successful catch-up around technological innovation is an important reason for their stepping into the newly industrialized countries. So it will greatly enlighten us to investigate these two countries.

2.2.1 The experience of Japanese independent innovation

Japan was severely damaged in World War II, but it rose rapidly from the ruins and became one of the major developed countries dominating the world's economic scene. How was Japan able to rise so quickly? In the perspective of the path towards the Japanese S&T development, compared with that of America at that time, its basic research was relatively weak and the original innovation fell far behind developed Western countries. But looking back at the Japanese history, we can easily discover: First, Japan had a good accumulation of innovation; second, after World War II, Japan put forward the policy of "building the country through technology," "building the country through patents," "building the country through S&T," and so on. Japan was oriented towards production development and comprehensive integration of products with commercial value. Thus, it promoted its process of industrialization and became a world power.

From the 1950s to the 1970s, Japan implemented the strategy of "introducing advanced technology from European countries and the United States and improving and innovating after imitating." Between 1950 and 1978, Japan spent $9 billion on signing a wide range of various technical contracts with the US and obtained a lot of fundamental modern technology from the US. This saved time and money for Japan to catch up and find a suitable path towards technological development. However, this strategy also caused a series of issues; especially the purchasing of a large number of foreign patents from the 1950s to the 1960s resulted in the Japanese industry being controlled by foreign technology. Thus, they had to keep purchasing patents from Europe and the US. It is not only expensive, but also with the fierce competition of international technology and the strengthening of the protection of IPR in many countries, it was not possible for Japan to continue the large-scale introduction of technology as in the 1950s. This seriously hindered the sustainable development of industry. Especially, when the industrial structure dominated by the iron and steel, the auto, and the machinery and chemical industries was transferred to one dominated by the electronic information industry, the situation of lacking necessary technological support appeared. Based on this, Japan began to gradually emphasize basic research rather than application research, and turned from technological imitation to technological innovation. The building of the "Tsukuba Science City" was one of the most important plans. In order to meet the needs of "building the country through S&T," the Japanese government moved more than 40 research institutes to Tsukuba Science City. This policy helped to form a concentration of academic research and education in national research institutes and universities, which promoted the research of major science projects.

In particular, a number of modern enterprises appeared in Japan in the 1950s. On the basis of introduction, implementation, and adoption, these enterprises transferred to improving and innovating in the 1960s. In the 1970s, Japan began to advocate a technology development organization of "cooperation among the government, industry, and university" in order to seize the opportunity of the new technological revolution. In the 1980s, Japan put forward the policy of "building the country through S&T," which sped up the process of independent innovation and gradually realized the transformation from imitation to independent innovation. However, at the same time, Japan did not exclude the introduction of cutting-edge technology from Europe and the US. According to the statistics from the Japanese government, the introduction of advanced technology increased by 86% from 1982 to 1986 in Japan, with a 103% growth in software and 69% in hardware; from 1986 to 1990, these three indexes were 91%, 153%, and 16%; from 1990 to 1992, they were reduced to 9%, 7%, and 2%. This was attributed to the gradual strengthening of the Japanese ability for independent innovation.

2.2.2 Japanese policy and measures of innovation

The gradual transformation from introduction and imitation to independent innovation in Japan had a strong relationship with the forceful policy support and guidance from the government.

First, the concentration of scientific research resources on the research of application development had both advantages and disadvantages. After World War II, the US made a large-scale investment in basic research. Based on the catching-up strategy, Japan put most of its research resources into application development and invested mainly in the field of engineering. In the early 1990s, in the investment of global basic scientific research, the US and European countries accounted for about 50% and 40%, respectively, but Japan only accounted for about 10%. Such practice sped up the process from introduction and imitation to innovation, but it has also favored hidden dangers of recession for the Japanese economy in the late 1990s and early 2000s. Because of the earlier contempt for basic research, technological breakthroughs and accumulation needed in order to continue independent innovation are insufficient now.

Second, the government gave proper protection to the research achievements of enterprises. Japan had a unique superiority in the incremental innovation dominated by process innovation, so the supportive intervention of the government in the technological development of the auto and electronic industries were very concrete after World War II. In order to encourage enterprises to carry out research on application development, the government gave preferential policy to patent applications, so that a patent was very accessible for an enterprise. At the same time, the protection of patents was relatively limited, which made enterprises use their competitors' innovation at low cost and risk. This institutional arrangement was extremely favorable to Japanese incremental innovation at that time.

Third, more attention was paid to the R&D of enterprises and the establishment of a cooperative innovation mechanism among enterprises was encouraged. The Japanese government paid more attention to promoting the establishment of an engineer's resource pool in large enterprises and put the emphasis on the development of knowledge, which is related to production, in large companies, so as to support R&D and innovation in enterprises. At the same time, when the enterprises promoted the development of products, the first step was to encourage the researchers to choose the research topic independently (47% of all research topics). Based on the emphasis given to the cooperation from the government, industry, and universities, the Japanese government actively promoted the establishment of an inter-firm cooperation mechanism in order to improve cooperative innovation together. The production process innovation called the "Toyota model of production" is an example of cost saving. The nucleus of the Toyota model is a demand-oriented dashboard manufacturing plan. Its advantages are low cost, high quality, and differentiation of production. In order to realize such an innovation, the government brought relevant enterprises together in positive collaboration.

Fourth, the Japanese government implemented the incorporation of national research institutes and national universities. Japan implemented the second phase of the "basic plan of science and technology" in 2001 and began the reform of national research institutes. The first was changing national research institutes into independent administrative corporations in order to enhance the flexibility of each institute in personnel and financial management; canceling the status of research institute employees as civil servants, therefore allowing research institutes to sign a contract of employment according to their own needs; and allowing research institutes to accept private funding (previously prohibited). The second was changing the way of making the working plan and financial budget program of the research institutes annually. The ministries and agencies in charge made regular assessments for the independent administrative institutions (IAI) , and determined the resource allocation of the government according to the results. In April of 2004, Japan began to implement the reform of the IAI of national universities. The first was changing national universities into IAIs, which were no longer a part of the Ministry of Education, Culture, Sports, Science and Technology (MEXT). But the quantity of the funding for national universities from the government had no significant change; the second was gradually reducing the number of national universities and merging some of them.

Fifth, the Japanese government improved relevant laws constantly and implemented the national S&T plan. For example, Japan officially announced their new Science and Technology Basic Law in 1995 and the "basic plan of science and technology" for a cycle of five years. The latter called for greater cooperation among industries, universities, and research institutes, which was to promote the establishment of new enterprises based on the "seed of technology" or on the originality of universities and research institutes. The Japanese government increased the number of postdoctoral scholarships, strengthened the support for young researchers, improved the mobility of researchers and allowed them to apply for research grants. Besides, there was an increase in competitive research funding so as to use it intensively, and an increase in the government's research resources. The second phase of "basic plan of science and technology" provided that competitive governmental research funding will double within 5 years.

Sixth, there was a push for the establishment of an organization system dominated by a horizontal flow of information, which is beneficial to innovation. In contrast to the vertical information flow of large companies in the US, Japanese enterprises preferred the horizontal flow of information, which improved the efficiency of internal information flow and decision making. This model was quite typical in manufacturing enterprises that had mass production. Japanese companies required employees to have more holistic knowledge and therefore they used the "job rotation system." They ensured the quality of products by lateral organization, where the quality management team absorbed various kinds of staff from different links. So in the comparative study of Japanese and US electronic enterprises, the management expert Baba gave the name "the whole factory is a laboratory" to the Japanese organizational structure with a horizontal information flow.

2.2.3 Typical cases of Japanese independent innovation

(1) Independent innovation of the Japanese auto industry

First, the history of the Japanese auto industry. The development of the Japanese auto industry has gone through four stages. The first stage was in the 1950s. At that time, the Japanese economy was in recovery. The auto industry was weak, with technology mainly imported, and its main production was trucks. The second stage was in the 1960s. This was the stage of scale development of the Japanese auto industry. After many years of borrowing foreign technology and equipment, the capability for independent innovation had improved, and they began to design independently and produce new models. Toyota, Nissan, and other enterprises began to produce at that time. The third stage was from the 1970s to the 1980s, a period of rapid development for the Japanese auto industry. Development speed, production efficiency, and export growth accelerated significantly, and exceeded that of the European and American auto industries. The fourth stage started in the 1990s and continues till now when the Japanese economy has been experiencing recession. Japan has begun to compete with other countries in the production of automobiles and in new product development. In particular, occupying a leading position in the technological development of new environment-friendly cars has laid the foundation for the sustainable development of the Japanese auto industry.

Second, the Japanese government's support for the independent development of the auto industry. After World War II, the Japanese government implemented the protection policy for the auto industry according to the "infant industry theory." The government officially announced a series of supportive regulations and policies. Only national capital could be invested in the auto industry, which cut off the competition between domestic and foreign enterprises, and promoted the rapid development of the Japanese auto industry. The government provided the enterprises with "support rather than intervention," which means there was no direct intervention in the enterprise production and operating activities, but support for the enterprises through relevant policies and regulations. Among the supportive policies, the first was the legislation protecting the development of domestic auto enterprises and resisting the input of foreign capital into the domestic market. The second was providing the enterprises with loans and preferential tax policies, so as to ensure independent innovation. The third was putting restrictions on auto importation by imposing high tariffs, and the incubation of independent auto brands. The fourth was encouraging enterprises to export in order to earn foreign exchanges, explore international markets, and become part of the international competition.

Third, the introduction of technology and independent innovation of Japanese auto enterprises. In the early postwar period, compared to European and American auto producers, the Japanese production technology was quite backward. To eliminate the gap in a short time and to reach the advanced level worldwide, the introduction of advanced technology from abroad seemed to be a shortcut. Therefore, from 1951 to 1969, Japan introduced 405 advanced technologies from the US, UK, Italy, and other countries. This exerted important effects on accelerating the development of the Japanese auto industry and promoting the study of auto technology. However, the introduction of advanced technology alone was not enough. Therefore, Japanese enterprises combined the advanced technology from European countries and the US with its technical innovations, which resulted in a rapid development of production and performance of the vehicle and the strengthening of international competitiveness. For example, Toyo bought the patent of the rotary engine from Wankel company of Germany in 1961, then researched and tested it first. After 200 hours of continuous experimentation, they found that the contact of the piston with the cylinder caused a problem of vibration. In order to solve this problem, Toyo spent $140 million on developing new materials. The problem was fixed successfully in 1964. The Japanese production and the quality of the rotary engine exceeded that of the German model. There are many similar examples in Japan. On this basis, Japan became the automobile producing country whose products are famous worldwide for their light weight, fuel-saving, and environment-friendly qualities.

(2) Independent innovation of the Japanese semiconductor industry

As an industry with cutting-edge technology and high added value, the semiconductor industry has an important strategic significance in the economy of any country; therefore, the developed countries regard it as the backbone industry. In the 1980s, Japan had become the center of the world's semiconductor industry. In Japan, the semiconductor industry policy is focused on promoting the application of the semiconductor in the field of industry and consumer products. In the early development of the semiconductor industry, in order to reverse the situation of over dependence on European countries and the United States, the Japanese government played an important guiding role in constructing a reasonable framework for orderly competition in the semiconductor industry.

First, in order to guide and promote the development of the semiconductor industry, Japan formulated the Law on Temporary Measures for the Promotion of Electronics Industry in 1957. The implementation of this law promoted the development of Japanese enterprises in the semiconductor industry on the basis of learning the advanced technology from the United States. Second, Japan formulated the Law on Temporary Measures for the Promotion of Specified Electronics Industries and Specified Machinery Industries in 1971. This law strengthened the development of the electronics industries dominated by the semiconductor. The implementation of this law helped Japanese enterprises to strengthen their own R&D and production capacity and to resist the impact from the European and American semiconductors effectively; this brought the Japanese semiconductor products onto the world market. Third, Japan formulated the Law on Temporary Measures for the Promotion of Specified Machinery and Information Industries in 1978. Its implementation strengthened the development of the information industry with the semiconductor as the core. Since then, Japan has not issued any further special regulations on the electronic information industry based on the semiconductor, but it has promoted comprehensive regulations with the purpose of promoting the development of the high-tech industry including semiconductors. representative of these was the legislation of the Science and Technology Basic Law in 1995.

The Japanese government also restricted foreign investment in the Japanese semiconductor industry and required it to purchase technology directly with the purpose of avoiding the control over the technology of the semiconductor industry by other countries. Until the 1970s, the Japanese government had always had a strict policy of industrial protection with tight control over imports. Regarding foreign investments in semiconductors in Japan, the Japanese government has always had a severe examination and approval system, including prohibiting the establishment of individual foreign proprietorship and requiring foreign enterprises to open all the patent technology to the joint venture. After 1975, these policies were relaxed, but the Japanese were still burning with ambitions in the field of micro-electronics. A typical example was the implementation of the very large-scale integrated projects (VLSI) put forward by the Minister of Economy, Trade and Industry (METI). The project was 40% financed by the Japanese government. Thus, the five largest semiconductor companies could carry out cooperative R&D of a new-generation product. At the same time, the Japan External Trade Organization (JETRO) provided Japanese semiconductor companies with the latest information through its branches in 55 countries around the world. Moreover, the Japanese government provided low-interest loans to semiconductor enterprises through the National Development Bank. The interest rate of the loan was close to zero, while on the US market, it was 4%–5%.

Over the years, Japanese semiconductor enterprises have always insisted on the principle of "self-design and self-production" and on establishing a coordinated process of the industrial system of design and production. In recent years, Japanese semiconductor enterprises have developed a new generation of semiconductor production technology by working with relevant manufacturers. The government occupies the dominant position in this process. The enterprises unify standards of equipment and technology, and concentrate on the development of system integration technology with high performance and low power consumption to meet the needs of high-performance components of small consumer electronics products. Major Japanese semiconductor companies hope to become the leading enterprises again in the global semiconductor industry through joint investment, cooperation, and alliance.

2.3 The Republic of Korea: Introduction and innovation

The Republic of Korea has become an important economic force in the world today. According to the report of the 2005 World Competitiveness Yearbook published by the International Institute for Management Development (IMD) in Switzerland, the national competitiveness and scientific competitiveness of the Republic of Korea ranked second and fifteenth in the world in 2004, respectively. Moreover, the Republic of Korea was the sixth or seventh power regarding IPR in the world. In the early 1960s, the Republic of Korea was a "poor" agricultural country and GDP per capita was less than $100. They could not even produce bicycles. But in 1962, after the implementation of the first five-year plan for economic growth, the Republic of Korea achieved rapid economic growth and created the "Miracle on the Han River." The Republic of Korea has been known as one of the "Four Asian Tigers" since then. In 1996, the Republic of Korea joined the Organization for Economic Co-operation and Development (OECD) as one of the main representatives of a new industrialized economy. The economic aggregate of the Republic of Korea reached the 11th position in the world in 2004 with GDP per capita over $14,000. The emphasis on independent innovation is an important reason for the Republic of Korea's making such remarkable achievements.

2.3.1 The characteristics of Korean independent innovation

(1) More attention to the actual effect of technology introduction

First, the introduction of technology into Korean enterprises has a clear purpose. Regarding the introduction of high-speed railway (HSR) technology, the Korean side clearly required the French side to transfer all HSR technology, including the technology of the vehicle, the contact line, the control device along with their operation maintenance technology, etc. In addition, the Republic of Korea had the right to use related IPR and to open up to the international market and produce 50% of the total cost. Second, the Republic of Korea paid attention to the introduction of non-trade technology, that is to obtain technology by sending personnel abroad to receive training, visit, and exchange technological information. An example is through the cooperative R&D between Samsung and Sony, many of Samsung's electronical products are more advanced than Sony's. Korean enterprises require the partners to train technical personnel, which, together with "learning by doing," rapidly improved the level of R&D ability of the Korean personnel.

(2) The implementation of the technology that has been introduced, with the help of technology alliances

The introduction of key industrial technology into the Republic of Korea has always been as follows: First, the government spends large amounts of money on introducing the most advanced foreign technology, and then organizes manpower and financial resources for scientific research. For example, the investment of the Korean government and local organizations in the CDMA reached 6 trillion wons, as did the introduction of high-speed rail technology. It laid a good technical foundation for independent innovation in the future. Especially, after the introduction of technology, Korean enterprises established a technology alliance, which was guided by the government through the invitation for bids, to tackle key problems jointly. The technology alliance avoided the situation in which a single enterprise might have no competence and gave up the project halfway. Like the CDMA project, the public-private partnership (PPP) and the internal and external collaboration were established. Thus, the Republic of Korea quickly mastered the core technology of CDMA, applied for a number of patents, and achieved a leading international position.

(3) Timely establishment of technology-oriented enterprises and research institutes in foreign countries

This helps to obtain foreign technology with the help of local technical forces. For example, major large Korean companies have established research institutes or branches and subsidiaries in the Silicon Valley in the US. These enterprises hire American engineers to design and develop by offering them higher wages.

2.3.2 The measures of Korean independent innovation

(1) The government's powerful S&T policy

The Republic of Korea is a small-sized country with restricted natural resources. Almost all of its energy is imported. Thus, the Korean people have a strong sense of crisis, just as a Korean pet phrase goes, "we must have the S&T to support the future survival of the Republic of Korea." Therefore, in the 1960s and the 1970s, the Republic of Korea introduced a lot of advanced technologies from the US and Japan, and paid more attention to implementation and absorption. For every $1 cost of technology introduction there would be an $8 matching fund in order to implement the technology. At the same time, the Korean government adopted a series of supporting measures concerning law, policy, funding, talent training, etc.

The first measure is legal protection. Since the 1960s, the Republic of Korea has officially announced the Science and Technology Promotion Act, the Science Education Act, the Technology Development Promotion Law, the Technology Assessment Law, etc. In 2001, the Republic of Korea announced the Framework Act of Science and Technology containing 29 kinds of laws, which covers all aspects of the S&T development, and provides legal guarantees for the S&T development as well as technological innovation.

The second measure regards clear strategies of S&T. After considering the situation, the Korean government launched appropriate technological strategies to advance with the times. In 1989, it issued the "five-year plan for the development of cutting-edge industry," which focused on the research and innovation of seven industries, including microelectronics, new materials, biological engineering, and optical fiber. In 1991, it launched the G7 project, lasting for 10 years. Its goal was to catch up with the S&T level of seven Western countries in the 21st century. In 2001, it made the Basic Plan of Science and Technology for five years (2002–2006) to develop the fields of 6T, namely IT (information technology), BT (biotechnology), NT (nanotechnology), ST (space technology), ET (environmental engineering technology), and CT (culture technology). In 2004, it launched the "ten new-generation growing power engineering projects," focusing on the development of ten high-tech industries, such as digital broadcasting, the intelligent robot, the new-generation semiconductor, and the future type of cars.

The third measure is to foster a rapid increase in investments in R&D. In 1962, investment in Korean R&D was 2.1 billion wons, only 0.28% of the GDP. It was only $320 million in

1980, 0.56% of the GDP. The investment in R&D has increased rapidly since 1980. It was $14.4 billion (2.53%) in 2002, $16.4 billion (2.64%) in 2003, and $19 billion (2.82%) in 2004. Now, the proportion of investment in Korean R&D as a share of the GDP is lower than that in Japan but much higher than that in the US, Germany, France, UK, and other developed countries.

The fourth measure is the preferential policy for the technology development in enterprises. The policy includes the reserve fund system of technology development, the fee and tax relief system of technology and talent development, the tax relief system of new technology promotion investment, etc. In order to ensure the implementation of its tax support policy, the Republic of Korea has modified the relevant regulations. Therefore, the technical development organizations (enterprises, institutes, etc.) have become the object of legal support, such as making full use of government procurement policy. The new regulation claims that government departments are allowed to make preferential procurement of domestic products even if their price is higher than similar foreign products and state-owned enterprises have to purchase domestic products preferentially. The government purchases the new technology developed by small and medium enterprises (SMEs) and subsidizes pilot-scale experiments and industrialization. The relevant laws and regulations claim that public authorities should purchase the products of SMEs according to their department budget and annual plans. In addition, there is a flexible use of tax policy. According to the type of enterprise, the reserve fund system of technology development is set up; related enterprises can extract the reserve fund for technological development at 3%, 4%, and 5% of their total income; and this fund can be included in the cost. There are also many other policies, for example, exemption from income tax for full-time researchers and from other taxes for the expenses of personal R&D, the deduction and exemption of turnover tax and income tax of new technological development, the exemption from income tax when transferring technology to local people or enterprises and half deduction of income tax when transferring to foreign enterprises, special tax preferences for venture investment, etc.

The fifth measure is an emphasis on scientific and technological talent. The Republic of Korea proposed to cultivate 10 universities to rank among the world's top 100 research universities before 2012; the Republic of Korea has introduced the system of "special national researcher" to cultivate 10,000 core researchers, focused on R&D of the tenth generation new power growth technology, and has had vast investment in core research. The national management system is set up for outstanding scientists who have the hope of receiving a Nobel Prize, offering them the chance to take charge of large-scale national research projects and obtain research funding. For this type of scientists, physical protection is provided, which is equivalent to "national senior level" of ministerial officials. The country spares no effort in the promotion and development of reserve talents in S&T. The Presidential Science Scholarship is set up to encourage outstanding senior high school students to apply for famous domestic and foreign universities in the attempt to cultivate world-class scientists. Domestic students can receive a scholarship of around 80,000 yuan during their university studies; the students admitted to a famous foreign university can receive a $50,000 scholarship. For winning in an international competition such as the Science Olympiad, the President meets the "scientific geniuses" who have won prizes in person, along with their parents and teachers. The winner can also be exempted from military service. For the investment of foreign high-tech enterprises, the government encourages them to set up research centers in the Republic of Korea. These centers employ Korean masters and doctors of science and engineering, while the government pays their salaries.

(2) Enterprises as the main force of independent innovation

Korean enterprises, especially the large ones, are the mainstay of independent innovation. Concerning investment in R&D, the Republic of Korea has experienced a transition from government leadership to enterprise leadership. The proportion of the investment by government and enterprises was 77% and 23% in 1970, 52% and 48% in 1980. The proportion of R&D investment in the top 30 listed Korean companies was an average of 3.15% of their sales revenue. Korean enterprises have begun to set up R&D institutions since the 1980s, which increased rapidly later. The number of these enterprises was 100 in 1983, 1,000 in 1991, more than 5,000 in 2000, and more than 10,000 in 2004. Some large companies even have more than one R&D institution. For example, Samsung has 42 research institutes, and the number of research personnel increased from 12,000 in 1997 (accounting for 22% of the total number of employees) to 24,000 in 2004 (36%). Employees with a master's degree or a doctorate increased from 5,000 in 2000 to 24,000 in 2004. R&D expenditures increased year by year, with 1.6 trillion wons in 1999, 2.4 trillion in 2001 and 3.7 trillion in 2003 (8.5% of the sales revenue). Hyundai has 14 research institutes, and the total R&D expenditure was 2.5 trillion wons in 2004 (5% of the sales revenue) and more than 3 trillion wons (6% of the sales revenue) in 2005. LG has 27 research institutes; the sum of its R&D expenditure was 2.4 trillion wons in 2004 and 3.4 trillion wons in 2005. Because of the focus on technological innovation, Samsung, Hyundai, and LG became members of the "100 most valuable brands in the world" in 2005 as investigated by the American magazine *Business Weekly* and global brand management consulting companies, among which Samsung ranked 20th, Hyundai ranked 84th, and LG ranked 97th.

(3) The Republic of Korea obtains numerous achievements from independent innovation, and quickly becomes a country with excellent scientific and technological abilities

According to the report of the Technical Standards Institute of the Ministry of Commerce, Industry and Energy (MOCIE), 22 kinds and 113 Korean technologies have been adopted as the international standard in the IT field alone. The 1997 Asian financial crisis damaged the Korean economy heavily, but the Republic of Korea was the first to get out of trouble. The Korean automobile, shipbuilding, semiconductor, and information technology (IT) industries started later than in China, Seoul's first subway began construction ten years after Beijing's, but now, the Republic of Korea has an overall leading position in these fields. It has become the largest shipbuilding country (40% of the world's shipbuilding market), and the sales share of mobile phones is 22% of the world's. Korean products are exported around the world. The technological innovation promotes Korean economic and social development, which in turn, promotes technological innovation. Now, the Republic of Korea has a high level of urbanization; the rural population accounts for only 7.5% of the total population. The income of farmers is equivalent to 70% of the income of city residents. The Gini coefficient is about 0.32. Koreans have a higher level of education nationwide (the college admission rate was 89% in 2004, which is said to be the highest in the world). In the United Nations' 2004 e-government appraisal, the Republic of Korea ranked fifth in the world and first in Asia. In 2003, the Korean comprehensive ranking in informationization was seventh in the world. The popularizing rate of the broadband network (ADSL) has reached 70% (first in the world), and 61 out of 100 people use the Internet (fourth in the world).

2.3.3 Typical cases of Korean independent innovation

Korean independent innovation has its own unique model. The first step is the introduction of key technology that is vital to the national economy and to people's livelihood in the industries with a high relationship with technology. With the support of the government, the technology is implemented and adopted until core technology has been mastered so that innovative activities can be carried out. Thus, the Republic of Korea has reached world class in key industries such as telecommunications, railway, electronics, etc.

(1) CDMA

CDMA technology was first developed by Qualcomm in the 20th century. It was the most advanced mobile communication technology in the world. But, it was not industrialized and commercialized in the United States. The future of this technology seemed risky and uncertain. However, in 1993 the Korean government introduced this technology based on a detailed investigation and a full demonstration, and it made a great effort to industrialize it. The government and enterprises made such an effort at implementation, adoption, and re-innovation that they achieved great success.

First, the government gave strong all-around support. The government as the policy maker played a dominant role in the process of developing the CDMA further. It listed the CDMA as one of the projects of "national policy," clearly putting forward the goal of creating a world-class industry and first-class products. Therefore, the government formulated and implemented a series of policies and measures for the development of the CDMA, including technological development, standardization, service operator license, etc. The government concentrated on manpower and material resources to support R&D institutions, and on enterprises to carry out technology implementation, adoption, and re-innovation. This policy guaranteed the further development of the CDMA.

Second, key problems are solved by PPP and internal & external collaboration. The Republic of Korea introduced only 7 out of 53 CDMA patent technologies developed by Qualcomm; the rest were developed independently. In the early development, the cooperative research between the Korean side and Qualcomm was engaged mainly in the design and development of the terminal and the MTSO. Meanwhile, with the guidance of an official institution, Electronics and Telecommunications Research Institute, Samsung, LG, and Hyundai worked cooperatively to tackle key problems. The united development system of public-private interacting and risk-sharing focused on the development of business application technology; this accelerated the implementation and adoption of imported technology, overcame technical difficulties, and cultivated creative talents. In the meanwhile, the minister of the National Intelligence Service often investigated and guided them face-to-face, so as to find and solve problems; the directors of research institutes participated in R&D together with the researchers which inspired their morale and guaranteed smooth development.

Third, strengthening the R&D organization, management, and coordination. The National Intelligence Service takes charge of and guides R&D activities; the project management group made up of the government, R&D institutions, and enterprises supervised the process and the quality of the R&D; the R&D personnel of TDX SPC exchange was absorbed to participate in the CDMA project, so that their experience and knack could be used in the R&D of the CDMA; production enterprises and service operators were also encouraged to participate in the project, which was able to combine R&D with production and service, resulting in a more rapid process of industrialization and commercialization; the competitive relationship between the project of the GSM terminal supervised by MOCIE made the CDMA project under the pressure of "who is better and faster," improved the efficiency of its development, and shortened its development cycle greatly. After the commercialization of the CDMA, the government introduced a competitive mechanism and selected five companies to develop commercial service. The government selected synchronous IMT-2000 operators proactively, which formed good conditions for the sustainable development of the CDMA industry.

Fourth, investment in capital and talent increases and the products of CDMA technology enter the international market. According to incomplete statistics, the investment of the Korean government and local organizations in the CDMA reaches 6 trillion wons. The number of R&D personnel is more than 1,000. The human input of the enterprises' R&D accounts for 15%–20% of the total. The government supports enterprises to open up foreign markets. It makes an effort to expand the exportation of its products. At the same time, the PPP promotes CDMA diplomacy, persuading other countries to adopt the CDMA, so as to expand the market. In overseas marketing, the Republic of Korea has gone from the exportation of simple mobile phone equipment to network design and commercial service operation.

(2) The high-speed railway (HSR)

The Republic of Korea started building an HSR through the introduction of French technology and put it into commercial operation in December 2004. It became the fifth country to have an HSR following Japan, France, Germany, and Spain. Korean HSR technology came mainly from the French company, Alstom. In order to maximize the creation of opportunities for domestic firms, the Republic of Korea only introduced the core technology for the vehicle system, instead of the road and power system technology. During the construction of the HSR, the Republic of Korea required the high-speed train system vendor, Alstom, to transfer all the technology, and provide a training plan, a maintenance manual, and the maintenance technology for the entire core system. The Korean Train Express (KTX) engineering symbolized the world's largest technology transfer (100%) of manufacturing technology. Local manufacturing accounted for 50%, the Republic of Korea mastered the full set of HSR system technology including research, production, use, maintenance of train tracks, catenaries, and the trains. The French National Rail Company spent 20 years to pioneer the process of the European HSR, but the process took just a few years in the Republic of Korea who claimed that the train technology was its own technology after the opening of the Gyeongbu Line and the Honam Line, and thought that the technology would no longer depend on foreign technology after 2007. By then, the whole train system would be independently manufactured by Korean technology; the country's next target is exporting the technology following the steps of Japan, France, and Germany.

First, signing the agreement under the conditions of innovation and market access. The Republic of Korea introduced HSR technology not only to meet its own needs, but also to target the international market. Therefore, achieving localization and opening up the international market was the primary condition for introducing the technology. In June 1994, the Republic of Korea and the French company, Alstom, signed a technology transfer contract with favorable conditions. It was explicitly stipulated that all the French HSR technology should be transferred, including the core technology and operation maintenance technology including the vehicles, the contact line, and train control devices; the contract also allowed the Republic of Korea to use related IPR, open up international markets, and realize the localization rate of over 50% at the total cost. This laid good foundation for Korean development of the HSR based on French technology. Therefore, on the basis of France's providing advanced technology, the Republic of Korea developed more advanced HSR technology through secondary development, and implanted it as an export-oriented industry.

Second, learning and mastering the core technology as soon as possible, so as to create the conditions for secondary development. After signing the contract, the Republic of Korea sent a large number of engineering and technical personnel to France to receive training and learn the system interface technologies of power, signal, construction, and track systematically. The Republic of Korea also sent equipment technicians to France to accept on-the-job training with technology transfer. The quality supervisors of the Korean Rail Network Authority (KR), which took charge of the Korean HSR project, went to France to participate in the supervision of the quality, test, and test run of the car manufacturing; in the meanwhile, the Korean community and Alstom designed a rail facility system together. In the four-year period from 1994 to 1998, the Republic of Korea has gained as many as hundreds of thousands of pages of technical data of the design, manufacture, and testing; the number of technicians accepting training in France and the Republic of Korea was about 2,000, which laid a good technical foundation for secondary development.

Third, developing Korean HSR technology. Completing the basic implementation and adoption of French technology, the Republic of Korea focused on developing Korean HSR system technology with the combined strength of domestic R&D institutions and enterprises. The government listed it into the "G7" plan (a high-tech plan aiming at catching up with the West and supported with capital and talent). 129 R&D institutions and enterprises participated in it. In order to realize the goal of reaching a localization rate of over 50% for the main local facilities (including train control devices of the power locomotives, equipment dedicated to passenger cars, contact line and automatic control system, interlock, CTC system, etc.), which was agreed upon in the contract, for the total of 44 train engine units, the Republic of Korea won the right for 34 train engine units to be produced domestically. Therefore, the localization plan and schedule were formulated. The KR was responsible for examination and approval, supervision, and accounting supervision. The enterprises failing to complete the domestic indicators would be punished by 20% of the total cost. French technicians were invited to stay in the factory to guide the production. The total number of French technicians that stayed in Korea was more than 1,000; they ensured the progress and quality of production as well as the achievement of the localization rate (55.4%) of the main system equipment and devices. At the same time, because of innovative R&D, a large capacity of induction motors and other new equipment was developed successfully, such as the high-speed train system with a top speed of 352 km per hour.

(3) The iron and steel industry (POSCO)

The development of the Korean iron and steel industry has experienced three stages: "copying imitation" in the 1970s, "creative imitation" in the 1980s, and "innovation" after the 1990s.

First, the introduction, implementation, and imitation of advanced foreign technology (the 1970s). The Korean iron and steel industry insisted on introduction in the early days. The Republic of Korea had no technology of its own, and had to learn from others; the purpose was to build factories and form capacity quickly, so as to meet the demand of the domestic market. Therefore, the Republic of Korea imported advanced steel equipment from Japan and sent some technicians to Japan's Nippon Steel & Sumitomo Metal and NKK to receive training. These technicians quickly mastered the equipment operation and production technology. At the same time, POSCO, formerly Pohang Iron and Steel Company, had also recruited Japanese experts to guide them. The mastering of automatic control and product quality technology solved the problems appearing at the beginning of the operation. Relying on Japanese equipment, technology, and their strong entrepreneurial spirit, No.1 Pohang Blast Furnace with a production of 1.03 million tons was completed in 1973, followed by No. 2 to No. 4, which made POSCO's annual capacity reach 8.5 million tons. And it became the 11th largest steel mill in the world; some technical indexes and the production efficiency exceeded that of many developed countries, except Japan. During the implementation and adoption of Japanese technology, Korean enterprises also made improvements on the equipment and process.

Second, the emphasis on the introduction of technology and independent development at the same time (the 1980s). POSCO's development aroused vigilance of developed countries; the Republic of Korea was confronted with a block on its technology import. Therefore, when Korean companies continued to purchase advanced foreign equipment selectively, they collected and analyzed paper data, sent a delegation abroad to study, carried out technical exchanges and personnel exchanges, and held technical seminars, all favoring the mastering of the intelligence and information of the world's advanced technology. At the same time, the Republic of Korea intensified its cultivating of independent development power, integrating resources and research as well as developing advanced technology. Therefore, the enterprise technology research institutes, the industry research institutes, and other research institutes, along with Pohang University of Science and Technology, were set up. These institutions formed the R&D and talent training system combining industry-university-research; the Republic of Korea strengthened its amateur training system, continuously encouraged study both at home and abroad; moreover, the country recruited a large number of senior technical personnel, and enhanced its enterprise R&D forces; the people devoted to R&D were rewarded handsomely. As a result, the Republic of Korea made the leap from a pure imitation stage to the stage of partial innovation in the field of iron and steel. The localization rate of equipment increased from 12.5% to 63.1%. The enterprise production scale expanded unceasingly, and the product price was the most competitive in the world. The quality of the hot rolling products reached its highest level ever.

Third, entering the new age of technological innovation (after the 1990s). In order to support the sustainable development of enterprises, POSCO accelerated its innovation and formulated the medium- and long-term technology development strategy and plan; the research focus transferred from solving the existing problems to meeting the future development demand; while continuous improvement of production efficiency and product quality went on, the adjustment of product structure, reduction of the cost, the development of the next generation of iron and steel technology and basic technology R&D were the focus. Therefore, POSCO devoted more energy to innovation: The first was a widespread collection, sorting, and analysis of foreign information; talents and funds sharply tilted towards the high technology R&D projects; internal and external resources were integrated to tackle key problems together. The second was to establish an efficient innovation system, to cultivate a large number of innovative talents and integrated technology development capability, and to achieve technological results with independent IPR. At the end of the 1990s, POSCO's production capacity exceeded that of Nippon Steel and became the world's largest steel enterprise; besides its price advantage, the quality and variety of its products had reached the level of Japan's. The third was the formulation of the "medium- and long-term technology development plan" in the 21st century, centering on traditional technology and strategic product technology, selecting 12 strategic projects for R&D, and focusing on developing special high-performance steel.

(4) Samsung

Samsung was the OEM of a small, black and white television made by the Japanese company Sanyo based on the "imitation strategy," which meant that Sanyo provided Samsung with product technology, design, and parts. Samsung was the production workshop of Sanyo for organizing production. Even in the early 1980s, Samsung had also depended on the imitation strategy to open the dynamic RAM (DRAM) market. But according to the technical agreement signed with foreign companies at that time, it was very difficult for Samsung to obtain basic assembly technology other than the key technologies; Samsung had to obtain this from various other channels on the one hand, while on the other hand, the implementation, adoption, and improvement of the key techniques came from within the company. Through imitative improvement, Samsung was finally able to produce black and white televisions that meet the demand of the Korean people. Following the television, Samsung developed the 14-inch color TV (1976), the microwave oven (1979), the home video recorder (1984), the IM DRAM (1986), the portable video machine (1989), etc. Since then, Samsung has begun to implement the strategy of "keeping up with the technology leader." After years of actively learning, adopting, and training, in the late 1980s and mid-1990s, the technology development ability and the technical level of Samsung's products had caught up with the world's most advanced companies. The gap had been narrowed significantly. In some fields, Samsung was close to or even equal to them, so was the speed of product launching. At that time, Samsung had made great achievements in electronics and had strong development ability.

Chairman of Samsung Lee Kun-hee put forward the grand blueprint of "making Samsung a world-class enterprise" in 1993. Samsung was eager to enter the high-end electronics market; the technology it needed was also more advanced. However, Samsung couldn't be completely independent of external companies, so it acquired and caught up with advanced technology in a variety of ways. On the one hand, it merged with other high-tech enterprises; on the other hand, it formed alliances with enterprises occupying the cutting-edge technology position and shared technology. As a result, in fewer than 40 years, Samsung climbed up the growth ladder of "imitator, follower, leader, and world-class enterprise." Its market expanded from low-end to high-end and spread out across the board. In the late 1990s, Samsung was fully able to guide the market demand with the most advanced technology, meet the high-end market demand, and win a plurality of firsts in the world. Samsung's success enlightens the industry in the following ways.

First, technology accumulation in the proper path of technology acquisition. i) Reverse breaks. Samsung imported components and parts with advanced technology which could not be produced by itself. After the introduction, not satisfied being just an OEM, Samsung began to decompose, learn, imitate, improve, and innovate. ii) Samsung employed foreign researchers of Korean origin to research and develop domestically. Every enterprise can do this at any stage, but this is especially suitable for the start-ups. iii) Samsung had the ability to introduce technology from small companies in advanced countries. Because the technical barriers of large companies were seriously high, the acquisition of high-end technology from them was impossible. Thus, purchasing advanced technologies from small companies that needed cash flow was a feasible solution.

Second, great expenditure on R&D and super government support. Samsung's turnover in 1997 was $15.3 billion, its gross profit $4.8 billion, but its investment in R&D was more than $1 billion every year. However, industrial investment depended on borrowing money, so that the debt to asset ratio was as high as 223% in 1997. At the same time, the government supported local key enterprises like Samsung and Hyundai with the strength of the whole country. The government provided guarantee for the huge loan of these enterprises. The numbers were several times the enterprises' net assets.

(5) Hyundai

The development of Hyundai has experienced five stages: i) The initial stage of assembly (1962–1966). The Kia, Hyundai, and other enterprises began to assemble cars using foreign technology. ii) The early stage of improvement and localization (1967–1971). The government focused on supporting the auto industry to enter into the stage of localization. The auto industry formed a distinctive production management model and an industrial development pattern. iii) The stage of developing domestic cars and preparing for exportation (1972–1976). In 1972, Hyundai began to invest independently in the development of domestic cars and the construction of the car factory with an annual production capacity of 80,000. In this stage, the localization level had a substantial increase, production technology went from simple assembly to independent development. iv) The stage of mass production and experimental exportation (1977–1981). The mass production system of the auto industry formed basically. Headed by Hyundai, each manufacturer also adjusted the strategic focus from the domestic market to the international market. v) The stage of consolidating the export base and promoting internationalization (1982–present). The Korean auto industry entered the international market on a large scale. The export quantity in 2004 was 2.3 million, the automobile has become Korea's leading export industry. The development of the Korean auto industry is a typical example with all the characteristics of Korean independent innovation.

First, introduction of technology and independent innovation into Korean auto enterprises. The technological innovation of Korean auto enterprises has experienced three stages. The importation of technology was the main content of the initial period of the development of the auto industry. In this stage, cars were mainly produced by SKD (semi-complete knock down) assembly. Then, the auto industry worked with Toyota and began to produce independently according to the CKD (complete knock down) method. In the 1970s, adoption and improvement was the theme. The government required that each company should select a vehicle model and develop the complete domestic car. In 1976, the localization rate of the major Korean auto enterprises was above 85%. At the same time, the auto production capacity increased (especially the production capacity of automobile parts). The auto enterprises spared no effort in cultivating their ability for independent innovation; Hyundai was a successful example. Its main products were cars and commercial vehicles. In 1974, Hyundai started to introduce the technologies of engine assembly, gearbox, and rear axle construction from the Japanese company Mitsubishi, while at the same time, they invited Italian companies to design the model and the body of the car. From the beginning of the 1980s, the Korean auto industry stepped into the independent development stage. In this stage, Korean car companies put product development in the most important position, emphasized the development of their own cars, built strong product development institutions, and trained developers. These enterprises invested a lot of development funds in cooperating with foreign enterprises. In 1986, the cost of R&D in Hyundai and Daewoo accounted for 4% of the sales and reached the level of advanced countries (3%–5%). Hyundai had more than 2,000 R&D personnel, and Daewoo about 500. The high input of production equipment and scientific R&D resulted in the development of various types of vehicles from the three companies of Hyundai, Daewoo, and Kia successfully, including various systems and all kinds of parts and components, such as the body, the chassis, the engine, etc., thus they entered into the stage of independent development.

Second, the support of the government for the independent development of the auto industry. In the process of developing the auto industry, the government provided the enterprises with powerful protection and support. Under relatively restricted conditions, the Korean auto industry achieved scale economy by mergers and rapid expansion. Through the introduction of technology and persistent localization, the Republic of Korea realized the leapfrogging development of industrial technology and developed an independent and completely national auto industry system. The government provided the enterprises and market with protection in many aspects. The government reduced the production cost of local cars and encouraged automobile exportation through a variety of subsidies and preferential measures. From 1962 to 1990, the Republic of Korea formulated a series of regulations and policies for fostering the development of the auto industry. At the same time, the government set up all kinds of tariff and non-tariff barriers to limit the importation of foreign cars, so that domestic auto enterprises were free from foreign competition.

2.4 China must insist on the national strategy of independent innovation

2.4.1 Independent innovation is beneficial to the sustainable development of the Chinese economy

Since the 1950s, several strategies have been put forward in China, such as "March to S&T," "S&T is the chief productive force," "Rejuvenate the country through S&T," and "Construct a national innovation system." These strategies contain the basic idea of independent innovation.

(1) The context for the formation of the independent innovation strategy in the new period

In China, the relevant concepts of independent innovation have experienced an upgrade from the integration into the S&T development plan to a national strategy. Especially in the 21st century, at the 18th collective study of the Political Bureau of the CPC Central

Committee on December 27, 2004, then General Secretary Hu Jintao pointed out that independent innovation should be put in a prominent position in all issues of S&T, so as to enhance the ability for S&T innovation and core competitiveness, and establish a road towards S&T innovation with Chinese characteristics during the practical process. We must aim at the forefront of world S&T development, speed up the construction of a national innovation system, and strengthen the ability for original innovation and integrated innovation. We should insist on the guideline "Doing certain things and refraining from doing other things," make use of those great tasks which have a strategic, fundamental, and crucial influence on the development of the Chinese economy, S&T, national defense, and society, and make a greater effort at bringing S&T resources into the field of strategic high technology related to the overall situation of modernization, into the field of public welfare research related to comprehensive, balanced, and sustainable development and into the field of basic research related to the sustainable development of S&T. We should intensify our tackling abilities of key problems for the breakthrough of science and technology. Some scholars think that this is a turning point in China's central policy-makers' understanding of independent innovation. The 21st, 22nd, and 23rd collective studies of the Bureau all regarded the improvement of China's capability for independent innovation as the core strategy.

In March 2005, at the State S&T Awarding Meeting, then Premier Wen Jiabao pointed out that independent innovation is the "muscles and bones" to support the rising of a nation. We must regard enhancing the ability of independent innovation as a national strategy, which should be carried out in various aspects of constructing a modernization plan for every industry and region. We should make our efforts at building an innovation-oriented country with international influence.

From August 19 to 23, 2005, then General Secretary Hu Jintao investigated into the provinces of Henan, Jiangxi, and Hubei again and had important discussions concerning the strategy of independent innovation. He emphasized that improving the capability for independent innovation was an important support for long-term stable and rapid economic development, for the adjustment of the economic structure and for a change in the economic growth model, as well as for the construction of a resource-saving and environment-friendly society, and for the improvement of the international competitiveness of the Chinese economy and its ability to resist risks. Enhancing the capability for independent innovation should be regarded as the focal point of technological development and the central link to adjusting the economic structure and changing the economic growth model, so as to establish a road towards S&T innovation with Chinese characteristics.

At the end of 2004 and the beginning of 2005, the Political Bureau of the CPC Central Committee held a meeting to discuss "the national medium-and long-term plan for the development of S&T," where the following guiding ideology was clearly put forward, "Independent innovation, leapfrogging major projects, supporting development, leading into the future with independent innovation as the core."

(2) Only with independent innovation can we realize sustainable development

The basic connotation of independent innovation strategy is to enhance the capability for independent innovation in China as the basis for improving national competitiveness, to take this as a great choice for China to be able to cope with future challenges, and to regard independent innovation as the strategic guiding line for the future development of S&T and the basic solution for the construction of an innovative country. From a historical perspective, only when we have a strong ability for independent innovation, can we hold the opportunity and win the initiative in the fierce international competition. In particular, in the key areas related to the national economy and security, core and key technology cannot be bought; we must choose independent innovation. Only when we master a number of core technologies in several important areas and own independent IPR, create a batch of enterprises with international competitiveness, and enhance national competitiveness, can we cope with the fierce competition on the global market and keep our enterprises in an invincible position, so as to promote the sustainable development of the Chinese economy.

In the mid and late 1990s, China entered the middle stage of industrialization, but the economic development mainly relied on increasing investment and cheap labor, and the high energy consumption of mineral resources. According to the International Monetary Fund's earlier published data, the Chinese GDP was $5.75 billion in 2010, the second place in the world, but only accounted for less than 10% of the world's total GDP. However, the oil consumption was 20% of the world's, iron and steel 20%, and cement 40%. If we do not change the situation as soon as possible, it will be difficult for the Chinese economy to develop sustainably. To get out of this embarrassing situation, what we can only do is increase the contribution of the total factor production (generalized technical progress) for economic growth.

During the 30 years of reform and opening up, depending on the introduction of large amounts of foreign capital, advanced technology and management experience, the Chinese economic development was effectively promoted. But the lack of core technology and independent IPR resulted in Chinese enterprises having more than 60% of foreign technology dependence. Thus, in order to increase the contribution of the total factor production (generalized technical progress) for economic growth, China will have to become dependent on the advanced technology of developed countries. But in the fierce international competition, the lack of core technology and independent IPR not only threatens national security, but also affects the promotion of national competitiveness and leaves China in a very disadvantageous position in the labor division of the international industrial chain. Based on these considerations, independent innovation must be taken as the basic strategy at present and in the future for the sustainable development of the Chinese economy. We must obtain core technology through independent innovation, cultivate independent brands, and grasp the control rights of the market and profits; we must speed up the actual process of the adjustment of the economic structure and the change in the model of economic development through independent innovation.

2.4.2 Related practice shows that underdeveloped countries are likely to realize independent innovation

The experience of China and other countries shows that, after a long period of hard work, it is possible for us to rapidly reach a good level of independent innovation and construct an innovative country from our present dependence on foreign technology and imitative innovation.

(1) The Republic of Korea and Japan have proved that underdeveloped countries can realize independent innovation

Among all industrialized countries, Japan and the Republic of Korea are emerging countries with an earlier industrialization process than China. The experience of these two countries has shown that it is possible for underdeveloped countries to realize independent innovation.

First, laying the foundation by imitation, and developing by improving innovation in Japan. From the analysis in Section 2.2, we have seen that Japan rose quickly from the ruins after World War II and became one of the major developed countries leading the world's economic trend. Relying on "building the country through technology," "building the country through patents," and "building the country through science and technology," and so on, Japan stepped rapidly into the process of industrialization, becoming a world power. And its core was the implementation of "the introduction of advanced technology from European countries and the US, innovative improvement after imitation." After making the effort from the 1950s to the 1980s, Japan found a suitable road towards its independent development. After importing a large number of foreign patents, the Japanese emphasis was gradually transferred from applied research to basic research, and then transferred from technology imitation to independent innovation. In this process, a group of enterprises gradually emerged. Accordingly, all of Japan's comprehensive S&T strength, economic strength, and innovation ability formed and got promoted. In the late 1980s, Japan had become an innovative country with strong independent innovation ability. In the period from 1982 to 1986, Japanese importations of cutting-edge technology increased by 86%, which decreased to 9% in the period from 1990 to 1992. Of course, the change from introduction and imitation to independent innovation was inseparable from the strong policy support and guidance from the Japanese government. For example, the Japanese emphasized investment in R&D, provided protection to enterprise achievements, and promoted the establishment of a cooperative inter-firm innovation mechanism. Besides, the Japanese government promoted the enterprisization of national research institutes and national universities, improved relevant laws and implemented the national S&T plan; moreover, it also emphasized the establishment of an enterprise organization system dominated by horizontal information flow, which is beneficial to innovation. All the above give a certain enlightenment to China.

Second, laying the foundation by introduction and then leading into the future by innovation in the Republic of Korea. From Section 2.3, it can be seen that the Republic of Korea was a "poor and empty" agricultural country in the early 1960s with GDP per capita less than $100, and it could not even produce bicycles. But in 2004, Korean national overall competitiveness and scientific competitiveness had ranked second and fifteenth respectively in the world, and it was the sixth or seventh IPR power. With the above achievements, the Republic of Korea gradually stepped into independent innovation from introduction and imitation. In particular, the introduction of technology focused on effectiveness. By means of technological alliances, the Republic of Korea implemented the technology that had been introduced and established by technological enterprises and research institutes in foreign countries at the right time. The government formulated a powerful S&T policy, and enterprises were the main force of innovation, and so on. These are the main reasons that the Republic of Korea was able to become an innovative country rapidly.

(2) Lots of Chinese enterprises have had a number of successful experiences of independent innovation

After the founding of the People's Republic of China, Chinese enterprises have experienced three waves of innovation. The first wave of innovation occurred in the traditional planned economy era when "independent innovation was in a closed state, with the government as the innovation subject"; the second innovation wave occurred after the reform and opening up when "technological introduction and re-innovation were in an open state," with the subject of innovation gradually changing from the government to the enterprise, and the position of the enterprise as the subject of innovation becoming more obvious day after day. After joining the WTO in the 21st century, with the chance for the adjustment of the global industrial structure, Chinese enterprises saw the third wave of innovation, which was characterized by "independent innovation in an open state." In this wave, many Chinese enterprises gradually established a dominant position by independent innovation.

Typically, in the field of agriculture, the team of hybrid rice led by the academician Yuan Longping, not only solved the food problem for 1.3 billion people, but this technology has also made the world's grain production increase. After joining the WTO, Daqing Oilfield, Huawei, ZTE, Haier, Chery, Lenovo, Nuctech, Datang, Neusoft, Zhenhua, etc. acquired a strong awareness of competition, survival, development, and innovation, and they realized that the survival and development of enterprises cannot come about without innovation. Moreover, they also realized that independent innovation must be market-oriented in order to strengthen the competitiveness of the enterprises' service, so as to carry out practical activities of independent innovation. Daqing Oilfield developed a stable oil exploration technology with high yield, which contributed to ensuring the Chinese oil supply security. In the iron and steel industry, the energy consumption level per ton of Baosteel, Ansteel, WISCO, and other companies has been close to the international advanced level through introduction, implementation, and re-innovation. In the machine tool industry, since the 12th Five-Year Plan, the high performance of the CNC machine tool technology has been greatly improved, and has broken the technology blockade of the US, Japan, and Germany. The manufacturing industry of communication equipment, represented by Huawei and ZTE, has made China world-class in communication equipment manufacturing through the strategy of introduction, implementation, and re-innovation. In the manufacturing industry of household appliances, the production capacity of Chinese electrical appliances has ranked first in the world, as it has the independent innovation of Haier, Hisense, etc. in recent years. In the transformation of the university S&T achievements, Nuctech is a typical example. The enterprise integrates the S&T achievements of Tsinghua University, and then developed independently. Its isotope detector technology of the van container is more advanced than its French counterparts. At the same time, the performance and price have many advantages compared to its Western competitors, which results in its success both domestically and abroad.

Particularly in the computer field with extensive use, regarding super computers, the R&D of the Galaxy and the Dawning led the US to reduce the limitation regarding price and quantity for exporting computers to China. Concerning basic software, domestic office software gradually opened up the market. In 2004, the govermental procurement of domestic office software exceeded that of Microsoft

Office; domestic office software was able to replace Microsoft Office in the general government offices. Relevant tests showed that the performance of domestic office software has reached 80% of that of Microsoft Office. Especially, for the R&D of the Office software of Microsoft Corp, about $7 billion has been invested for more than 15 years. However, the R&D investment of Chinese enterprises is less than 100 million yuan, and the R&D cycle is much shorter. In terms of operating systems, several years ago, China had to buy foreign operating systems which are worth billions of yuan. But in recent years, China has independently developed an operating system based on the open-source Linux, and it has gradually mastered the technology of the operating system. The domestic Linux server has been widely applied in many areas, and the domestic desktop Linux has been used in some government departments as well.

2.4.3 China has the basic conditions and environment to implement independent innovation

The development of Chinese S&T and economy during the over 30 years of the reform and opening up vividly interprets great experience on the road towards independent innovation with Chinese characteristics.

(1) The central government made a series of important decisions and strategic deployments for the development of Chinese S&T

The experience of the development of S&T and the economy after the reform and opening up has proven that the central government should always have a broad and long-term view in the important moments of national development and make appropriate strategic deployments on the development of S&T and the economy, including the exploration of the independent innovation theory and pre-deployment.

First, establishing the guiding ideology of "S&T is the chief productivity" and promoting S&T to serve economic and social development. In March 1978, the Central Committee of the CPC held the First National Conference on Science and Technology; in his speech, Deng Xiaoping reiterated the basic Marxist viewpoint that "S&T is productivity." In order to meet the needs of reform and opening up and economic construction, the Central Committee put forward the major principle that "economic construction must rely on S&T, and S&T must serve the economic construction." In 1985, the Central Committee made "the decision to reform the S&T system" and clearly claimed that S&T achievements must be widely applied in production. In 1986, the Central Committee set the goal of "developing high technology" and "realizing industrialization," and launched the high technology R&D program (863 Program) and the Torch Program aiming at the industrialization of high-tech achievements. In 1988, Deng Xiaoping put forward the scientific statement "S&T is the chief productivity," which laid the early ideological foundation for the S&T development and the present strategy of independent innovation.

Second, the implementation of the strategy "revitalizing the nation through science and education," and the deployment of the development of S&T at the national strategic level. Independent innovation must be based on independent research and development and on personnel training. In the early 1990s, in order to ensure a smooth realization with modernization with the three-step development strategy, the CPC National Congress held a national S&T conference in 1995; it issued "the decision about accelerating the progress of S&T," clearly put forward the idea of "putting S&T progress and innovation in the key position of economic and social development"; moreover, it announced "the strategy of rejuvenating the country through science and education" and "the strategy of sustainable development as well as transferring economic construction to be dependent on S&T progress and the improvement of the labor quality." In 1997, China began to implement the National

Basic Research Program (973 Program), so as to solve the key scientific problems in economic and social development in the deeper and more extensive fields. In 1999, the Central Committee and the government held the First National Technology Innovation Conference and issued "the decision on the strengthening of technological innovation, developing high-level technology, and realizing industrialization"; it also proposed to further optimize the layout of S&T power and the allocation of S&T resources, to formulate and improve relevant policies, to promote the transformation of applied scientific research institutes and to update the classification of public research institutes, and to improve the support ability, public service ability, and sustainable innovation ability of S&T.

Third, enhancing the independent innovation ability and building an innovation-oriented country. After entering the 21st century, technical innovation further became the focus of international competition. In January 2006, the Central Committee held the National Conference on Science and Technology again; it promulgated the "national medium- and long-term science and technology development plan (2006–2020)," and issued "the decision to implement the layout plan of S&T and enhance the ability for independent innovation." This conference made the major strategic decision to strengthen independent innovation and build an innovative country, and put forward the guidelines of "independent innovation in key areas to support development and lead in the future." The 17th CPC National Congress made new requests: "significant enhancement of the independent innovation ability, a sharp increase in the amount of contribution of S&T progress to economic growth, and becoming one of the innovative countries," in order to realize the goal of constructing a society that is well-off in all aspects. Enhancing the independent innovation ability and building an innovation-oriented country are regarded as the core aspects of national development strategies and the key to improving comprehensive national strength.

(2) Substantial improvement in technical conditions and the institutional environment that are beneficial to independent innovation

During over 30 years of reform and opening up, there were profound changes from the concept of scientific research to the position of S&T, from an S&T system and mechanism to environmental research conditions, and from scientific layout to the strength of S&T, which resulted in a rapid increase in S&T activities and a sharp improvement in the level of S&T achievement.

First, basic conditions of scientific research have been improved significantly. Basic conditions have been completed for scientific research, including state key laboratories, large scientific instruments, natural S&T resources, scientific data, S&T literature, etc. By the end of 2009, there had been 3.184 million R&D and technical personnel, ranking first in the world. From the beginning of 2003, together with relevant departments and regional governments, the Ministry of Science and Technology and the Ministry of Finance started "The platform construction plan for the basic conditions of S&T." The digital technology platform was built based on the informationalization of the conditions of S&T resources; the network platform was also built to serve the whole society; the mechanism and policy were improved for the opening and sharing and performance evaluation; the construction of a standard system was promoted for complete scientific data and data sharing. These promoted an open sharing and the effective use of scientific resources. At the same time, the scale of investment in S&T kept increasing. In 2007, the total expenditure in social development funds was 366.4 billion yuan, which accounted for 1.49% of the GDP.

Second, basic research has achieved great original innovation. From 1996 to 2010, the number of Chinese SCI papers went from the 14th place to the first in the world. In 2006, the sum of Chinese patent applications ranked fourth in the world. In some great technological fields, China has achieved breakthroughs and gained a number of great independent innovation achievements, representative of which are the manned spaceflight and lunar exploration projects, the super hybrid rice project, high performance computers, and ultra large-scale integrated circuits.

Third, the ability of S&T to support industrial technological innovation has significantly improved. The technological innovation capability of basic industry, of the manufacturing industry and of new industries has enhanced markedly; the ability for independent development of a great technological and technical level of complete sets of equipment has obtained significant improvement, which has supported the engineering construction of the Three Gorges Project and the Qinghai-Tibet Railway. The ability of S&T in trying to solve the problems of saving energy, reducing emissions, and climate change has also improved significantly.

Fourth, the institutional environment stimulating independent innovation keeps optimizing. The laws and regulations based on the Law of Science and Technology Progress have improved sustainably; more than 30 systematical policies and measures meeting those introduced by the "national medium- and long-term science and technology development plan (2006–2020)" were formulated and helped to initially form a legal environment conducive to independent innovation. In particular, the environment related to IPR of independent innovation was greatly improved. For example, a relatively complete system of intellectual property was established. A series of laws and regulations were formulated, such as the Patent Law, the Trademark Law, the Copyright Law, the Countering Unfair Competition Law, Regulations on the Protection of Computer Software, Regulations on the Protection of Layout Design of the Integrated Circuit, and Regulations on the Protection of New Varieties of Plants. We have basically achieved this by acting on international convention. A relatively complete IPR protection system was built within a period of more than 20 years in China, while within a period of hundreds of years for Western countries. Accordingly, from 1999 to 2005, the number of patent applications to the Chinese Academy of Sciences was 15,092, which was 4.5 times the sum of the previous 14 years. Since 1998, the number of patent applications in the field of national defense has also shown a rapid growth trend with an average annual increase of 46%. At the same time, the breadth and depth of international S&T cooperation combined with Chinese independent innovation has expanded further. China has established a relationship of cooperation in S&T with 152 countries and regions, and it has participated in the International Thermonuclear Experimental Reactor (ITER) project, the Galileo Global Satellite Navigation System (GNSS), the Human Genome Project (HGP), and other major international science and engineering projects.

2.4.4 While insisting on independent innovation strategy, attention should be paid to its relevant restraining factors

Independent innovation is helpful to the sustainable development of the Chinese economy. The experience of China and other countries over several years has proved that it is entirely possible for an underdeveloped country to realize independent innovation. China has created the basic conditions for environment construction to be conducive to independent innovation. However, the Chinese independent innovation strategy also faces a series of restraining factors.

(1) Technical foundation and ability

Since the reform and opening up, the technical level of lots of industries in China has been improved to different degrees; some industries also have a certain extent of ability for technical competition on the international market. But overall, the technological foundation of many industries is still very weak, and technical ability very poor, which is bound to restrict the process of Chinese independent innovation.

First, the ability for introduction, implementation, and re-innovation is weak. Of the three types of independent innovation, original innovation is the most difficult; integrated innovation depends on a new effort; and implementation, adoption, and re-innovation is the most realistic. Currently, the ability for implementation, adoption, and re-innovation of imported technology is weak. Implementation and adoption is essentially a learning process of advanced technology and an approach to the accumulation and formation of the technical ability of underdeveloped countries in the process of catching up with advanced countries. There are some reasons why the ability of introduction, implementation, adoption, and re-innovation is weak. The first is the little attention that has been paid to the implementation and adoption of imported technology for a long time, or introduction without implementation and adoption. For example, the proportion of the money spent on importing to investment in the implementation and adoption of the technology that has been introduced into Chinese industrial large and medium enterprises (LMEs) accounted for only 1 : 0.15 in 2004. In other words, when \$1 was spent on importing, only \$0.15 was used for implementation and adoption, while those proportions in Japan and the Republic of Korea were 1: 5 and 1: 8, respectively. The second is the lack of technical ability for the adoption and implementation of imported technology. Nowadays, the majority of Chinese industries are situated at the end of the international industrial value chain, and the overall technical level is relatively low. Regarding the shipbuilding industry, China is still in a backward situation in the key technology field. In ship designing with high added value, there is a big gap between China and other countries. Compared to Japan and the Republic of Korea, the Chinese shipbuilding industry falls behind by about 10 years and behind the overall level of the ship equipment industry by about 20 years. Other industries also have this problem. Because of the current low level of technology, it is very difficult to implement and adopt advanced foreign technology in these areas.

Second, the situation of dependence on advanced foreign technology will be difficult to change in the near future. China has implemented the strategy "market for technology" over a very long period, but has not obtained core technology. And the enterprises in developed countries still pose a technology blockade on emerging countries by means of a patent strategy. First, though China is a country that exports industrial products, Chinese R&D ability for advanced processing technology and major technical equipment, which reflects industrial competitiveness, is very weak; coupled with a backward product design, it all leads to a crucial shortage of technical products with high technology and high added value in a market with an urgent need of them. At the same time, the acquisition of some important equipment required by the country and high technology industries depends on importations. For example, 100% of the optical fiber manufacturing equipment, 85% of the integrated circuit manufacturing equipment, 80% of the petroleum chemical equipment, as well as 70% of car manufacturing, numerical control machine tools, textile machinery, and printing equipment depend on importations. The more important and higher technical value-added the equipment is, the more importing and the lower the self-sufficiency rate will be. In general, industrial equipment with excellent technical and economic performance accounts for only 1/3, and nearly 1/5 has already aged and its overuse rate is nearly 40%. Second, under the technology blockade of developed countries, the acquisition of key technology depends on foreign enterprises. There are few Chinese independent IPR in industrial technology. Among the authorized invention patents in China, the foreign enterprise patents accounts for 2/3, most of them in the high-tech industry. At the same time, in the high-tech field worldwide, the United States and Japan have about 90% of all the patents in the world; other countries, including China, only about 10%. The lack of independent IPR will inevitably lead to the purchasing of foreign patents in order to acquire key technology to develop some products or industries.

Third, the technology blockade of developed countries is intensified, which can be divided into three categories. The first is a wide range of patent layout. Taking the auto industry as an example, from September 1985 to November 2005, 10 major domestic auto enterprises, including FAW, Dongfeng, Chery, and Geely, acquired 116 patents accumulatively. Fourteen major joint ventures, such as Shanghai Volkswagen, Shanghai GM, and Dongfeng Nissan, acquired 10 patents. In the meanwhile, Honda alone acquired 2,654 patents, 20 times the sum of Chinese enterprises; in addition, the Japanese company Toyota acquired 960 patents, 580 for Japanese Nissan , 402 for Korean Hyundai Motor, and 200 for German Volkswagen, much higher than that of Chinese domestic auto manufacturing enterprises. Under the background of the patent layout of the enterprises of developed countries, the rising Chinese auto industry inevitably faces numerous intellectual property barriers.

The second is transferring the "patent pool" to the industry standard. In some technological fields, some enterprises in developed countries combine core patents in a "patent-pool," and then establish the industry's technical standards based on it. They leverage their economic and technological advantages to control the international standard organization on the one hand; on the other hand, they construct patent-pool based on their own patents, and then formulate standards with their national power, so as to make their patent-pool become international industry standards in some fields, such as WCDMA and CDMA2000, etc. Therefore, the developed countries have established an insurmountable monopoly in some industries and control the industrial technology development track of some newly-industrialized countries. The enterprises in underdeveloped countries can only take orders from developed countries. So are some Chinese industries such as the computer, the mobile communications, the integrated circuit, the auto manufacturing, the numerical control machine industries, etc.

The third is turning the basic research achievements into patents. In 1980, the United States issued the Bayh-Dole Act, which broke the long-standing international convention that "the achievements in scientific research with federal funding will be turned into public products" and allowed the achievements of scientific research funded by public resources to be translated into patents of the funding institution. Other developed countries followed the United States and modified their patent laws, which resulted in basic research achievements falling into the field of private goods. Regarding the field of biotechnology, the United States and other developed countries allowed researchers to translate the functional gene, gene isolation tool, and its possible future use into patents, for example the whole research chain of pure protein, gene isolation, gene expression sequence, gene isolation reagent, semi-antigen, and vaccine, has formed a patent. Under this rule, the discovery of a functional gene means the occupation of a subdivision industrial field. Once they have occupied an industrial field, other countries must be subject to their patents in the process of technological development and technology path. In terms of this, the Chinese biological industry is suffering.

(2) Fund input and personnel

From the experience of Japan, the Republic of Korea, and China, enough funds and personnel input are a must to transfer imitation innovation to independent innovation. However, at present, China faces insufficient fund investment and personnel reserve for independent innovation.

First, investment in technological innovation is seriously insufficient. Innovation investment reflects the degree of attention a nation pays to technological innovation, but Chinese innovation input has remained far lower than developed countries and even lower than some other developing countries. For example, as for the proportion of R&D investment in the GDP from 1999 to 2003, it was 3.035% in Japan, 2.494% in Germany, 2.672% in the US, 2.215% in France,

1.908% in Canada, 1.863% in Britain, 1.54% and in Australia, while in China it was only 1.086%. But at the same time, the proportion of S&T funding in fiscal expenditure showed a downward trend, falling from 4.43% in 1995 to 4.12% in 2008. Due to insufficient government investment, innovation investment of enterprises was even less sufficient. The proportion of R&D funds in the sales revenue of industrial LMEs has been 0.8% for a long time. However, this proportion in developed countries was 2.5%–4%. The proportion of S&T expenditure in the main business revenue in China was 1.19% in 1995, and 1.57% in 2008, still lower than most developed countries of more than 5%. The problem lies in the fact that the enterprises with more than 5% of the main business revenue spent on S&T have enough competitiveness; the enterprises with 2% can only maintain their development, and the enterprises with less than 1% are finding it difficult to survive from a general international view.

Second, generally, enterprises lack talents and corresponding mechanisms suitable for independent innovation. Talents are the key to enhancing the independent innovation ability of enterprises. Taking an anonymous enterprise as an example, it recruited more than 20 R&D personnel from an enterprise in northeast China to develop and produce large- and medium-sized woodworking machines in 2001. In the same year, the enterprise independently developed and produced the BSG2312E type sander with three shelves and BPF2225F type joinery felt-board machine with high technology content, and pushed them onto the market at the right time. That alone made the enterprise successfully achieve a sales revenue of 180 million yuan in the same year. At present, its woodworking machine has been developed into six categories with 162 varieties, its corporate profit increased sharply, and export volume doubled year after year. It is not difficult to see the importance of talents for independent innovation. But the problem lies in the following aspects: The first one is the lack of suitable talents for independent innovation in the majority of Chinese enterprises. This has led to the low proportion of industrial LMEs with special research institutes. In all industrial LMEs, 39.8% of them had special research institutions in 1995, but it decreased to 24.66% in 2008; in all industrial LMEs, they had an average of 0.569 scientific research institutes in 1995, but it was even lower at 0.328 in 2008. The second aspect is the low proportion of top talents who are willing to work in an enterprise that lacks independent innovation. According to a relevant survey, more than 80% of the engineering students are willing to get a job in universities and colleges, scientific research institutes, and transnational corporations, but the proportion of students hoping to work in domestic enterprises is very low. In order to enter transnational corporations, some students prefer to wait and remain unemployed. For example, the number of engineering graduates in 2003 was more than 40,000, including about 1,000 engineering PhD graduates. However, the number of personnel engaged in technology development with a PhD degree in more than 2,700 large-scale enterprises was only more than 3,000. The third is the lack of a mechanism in lots of enterprises in making the talents play a role in independent innovation, which to some extent suppresses the enthusiasm and creativity of technicians in technological innovation.

A weak technological foundation and ability and insufficient capital and talent investment limit the process from imitative innovation to independent innovation. At the same time, it deserves to be mentioned and emphasized that the limits in thinking, policy, and culture puts a restriction on the independent innovation.

Chapter 3

The efficient path towards Chinese independent innovation: Understanding through case studies

Independent innovation is beneficial to the sustainable development of the Chinese economy. The experience of other countries and China itself over the years shows that it is entirely possible for underdeveloped countries to move forward on the road towards independent innovation. China has had the basic conditions and environment conducive to independent innovation, but in order to move forward on the road to independent innovation, an efficient path towards independent innovation is yet to be explored. This chapter explores an efficient path based on case studies.

3.1 Case selection

3.1.1 Basic considerations and the principles of case selection

This chapter discusses the efficient path towards independent innovation in China and begins with case analysis. Given the diversity of enterprises, this chapter will consider 8 factors to select cases: The difference in the nature of the enterprise, the innovation model, innovation paradigm, the innovator's background, the nature of the industry, the process of innovation, the nature of the product, and the results of innovation (shown in Table 3-1).

Table 3-1 Basic considerations of case selection

Case feature	Specific characteristics			
Nature of the enterprise	State-owned enterprise	Private enterprise	Joint venture	University enterprise
Innovation model	Original innovation	Imitative innovation	Cooperative innovation	Integrated innovation
Innovation paradigm	The regular innovation of mature enterprise		Unconventional innovation of pioneering enterprise	
Innovator's background	Overseas education		Domestic education	
Nature of the industry	Emerging industry		Traditional industry	
The process of innovation	Product innovation and breakthrough innovation		Process innovation and incremental innovation	
Nature of the product	Assembled product		Non-assembled product	
Result of innovation	Success		Failure	

Based on the above considerations, the case selection in this chapter follows these principles:

First, the case of the innovation model with typical characteristics will be selected. We try to choose some cases that are representative of some aspects. An example is Hisense, which is a state-owned enterprise in a traditional industry. Its innovation model is integrated innovation, and the innovation paradigm is regular innovation of mature enterprises; Xunlei, Tencent, and Chipscreen are private enterprises in emerging industries, whose model of innovation is independent development, and their innovation paradigm is unconventional innovation of pioneering enterprises; Chint is a private manufacturing enterprise in Wenzhou and it is an innovation model in itself. However, it is hard to regard one case as a category by a single standard, so the case can only be tentatively classified according to its typical characteristics.

Second, the enterprises in the Special Economic Zone (SEZ) of Shenzhen will be selected as the cases of regional industrial innovation. Shenzhen is at the forefront of the reform and opening up; after more than 30 years of reform and opening up, technological enterprises have mushroomed. Since the implementation of the independent innovation strategy, these enterprises have also taken the lead in staying ahead in independent innovation, and it is not hard to find that they have made achievements along diversified innovation paths regarding all of the characteristics of innovation. There are two types of independent innovation in Shenzhen. The first is the independent innovation model oriented towards the market from low-end product innovation to high-end product innovation. The representative enterprises of this type of innovation are Huawei, ZTE, BYD, Mindray, and CIMC. After transition and acceleration of this innovation model, these enterprises have steadily progressed along the road towards independent innovation. The second type is the emerging enterprises, like Xunlei, Tencent, Chipscreen, etc. These companies started relatively late, but directly from high-end R&D supported by venture investment. Therefore, the typical "knowledge and technology-oriented" independent innovation model formed. Therefore, Shenzhen as a model of regional independent innovation is very representative.

Third, some representative industries will be selected as industry analysis cases. The communications equipment manufacturing industry has developed rapidly in recent years. It is a typical independent innovation industry; it is necessary to study the typical cases in this industry. China still has some long-time backward industries, which have depended on the introduction of technology to survive over the past years; an example is the equipment manufacturing industry. In recent years, the equipment manufacturing industry has been developing rapidly, for example, Sany Heavy, Weichai Power, Foton Motor, and aviation products are very representative. At the same time, with the shortage of traditional energy, such as oil, coal, and natural gas, many countries have focused their attention on the development of renewable energy like wind, water, and solar energy. Although China started late in this area, there are some companies which depend on independent innovation to rank at the top in the world in recent years, so an analysis of the innovation carried out by these enterprises also has significance.

Fourth, the enterprises with the characteristics of regional development will be selected as typical enterprise cases. At present, except Shenzhen, the cities of regional development with typical significance are Wenzhou and Suzhou. The former model lacks innovation and was based on imitation, but Chint has stepped into the international market through "introduction, imitation, and re-innovation." The early Suzhou model did not have much technical innovation, but there has been plenty of innovation in the field of emerging industry in recent years, such as Suntech Power, which has become an internationally famous enterprise in this field thanks to its independent innovation. Similar enterprises also have important research value.

In view of the above considerations, the case enterprises are mainly distributed in Shenzhen (7), Beijing and Tianjin (4), Shandong (3), Jiangsu and Zhejiang (3), Anhui and Hubei, etc., the numbers in brackets represent the number of cases selected.

3.1.2 Brief information concerning the case enterprises and the case list

The cases under analysis in this chapter are shown in Table 3-2.

Table 3-2 Enterprise cases

Number	Cases	
1	Huawei	Independent innovation as the challenger's passport
2	ZTE	Independent innovation serving the world
3	BYD	From a battery king to an auto manufacturer

Number	Cases	
4	CIMC	Independent innovation forging a world-class enterprise
5	Xunlei	A dominator and pioneer in the downloading engine industry
6	Tencent	Winning by "Q"
7	Mindray	The leader of the Chinese medical equipment industry
8	Chipscreen	Leading the innovation in Chinese drug development
9	Foton	The innovation of independent innovation model—integrating knowledge and united innovation
10	Hisense	Chinese TV's "confidence"
11	Weichai	Independent innovation forging powerful "Chinese heart"
12	Suntech	Chasing the sun with new technology
13	CHINT	From a Wenzhou manufacturer to a world-class brand
14	FAW	Independent innovation creates "national car"
15	Chery	Creating the independent brand of the Chinese auto industry
16	Sany	A black horse in the Chinese equipment manufacturing industry

3.2 The path towards independent innovation from low-end to high-end (implementation, adoption, and re-innovation)

The process of introduction, implementation, adoption, and re-innovation is one of the roads to independent innovation that are followed in order to realize innovation from low-end to high-end in our country. As an underdeveloped country, there is an enormous gap between China's technical level and that of developed countries. Based on the late-developing advantage theory, the technological catching-up theory, and the technological leapfrogging theory, introducing advanced technology from developed countries is an important approach for developing countries to achieve rapid catching-up.

3.2.1 The connotation of introduction, implementation, adoption, and re-innovation

The introduction of technology and re-innovation can

also be called "secondary innovation." An academic interpretation of the concept "secondary innovation" was explained with the help of the "technological paradigm" theory of the British scholar Dosi (1982). The theory affirms that the technology paradigm is a totality of principles, regulations, methods, standards, and conventions, which are used to solve specific problems, and it is adhered to by designers, engineers, and managers. It stipulates the direction and the connotation of technological progress. In fact, every major innovation means defining a new technological paradigm.

Secondary innovation comes from imported technology, so it is essentially a series of improvements on the performance of the product and the manufacturing process along the established technological trajectory. The improvement is influenced by the level of scientific research and the market demand in the place of the innovation (Wu Xiaobao & Ni Yifang, 2001). Sun Bing & Shang Qingyan (2002) think that secondary innovation is based on the importation of technology, and it is constrained by existing technical regulations and develops along the established technological trajectory. Secondary innovation is beneficial to reducing the costs of R&D and avoiding repeated low-level R&D; moreover, it is advantageous for the late-developing countries to carry out independent innovation on the shoulder of the "giant"; it is also advantageous for technical accumulation and provides the basis for future original innovation.

It is necessary to illustrate that the introduction and exportation of technology was generalized as technology import and export and defined as "overseas to domestic or domestic to overseas" in the Technology Import and Export Management Regulations of the People's Republic of China issued by the government in 2001. "Transfer technology through trade, investment, or economical and technological cooperation" includes the transfer of patent right, patent application right, patent licensing, technological know-how, technological service, and other forms of technology transfer; the technology import is equal to the technology introduction. The concrete ways to import technology are purchasing technology, purchasing equipment and production lines, technology consulting, technology services, trademark rights, etc. In the process of introducing technology, enterprises should first focus on the advancement of the technology that has been introduced; second, enterprises should pay attention to the effectiveness of the imported technology, and reducing the importation of repetitive technology and improving the efficiency of the technology introduction. In the period from 1979 to 1998, 27,829 technologies were imported into China with a total contract value of $105.48 billion; from 1999 to 2005, nearly 50,000 technologies were imported into China, with the total contract value more than $100 billion, including technical fees amounting to $62.3 billion, accounting for 62.3% of the contract value. From 2006 to 2008, 30,481 technologies were imported into China, with the total contract value $74.57 billion, including $57.709 billion of technology fees, accounting for 77.38% of the contract value. The proportion of purchasing technology costs has increased gradually.

3.2.2 The evolution process from low-end to high-end

Secondary innovation is an important way for late-developing countries to achieve technological catching-up. Foreign studies on secondary innovation have focused on the process of technological catching-up. The most representative is the Korean scholar Kim Linsu (1997), who puts forward the "three-stage model." He puts forward the analysis framework of the global technological environment; shows two different innovation paths between developed countries and late-developing countries; and proposes the innovation path of "introduction—implementation and adoption—innovation" by taking the Republic of Korea as an example. Since then, Lee & Lim (2001) have put forward the "stage-skipping theory" based on the "three-stage model" of Kim Linsu (1997). They affirm that late-developing countries do not have to follow the path of the pioneers in the catching-up process. They are likely to skip certain stages and achieve catching-up more rapidly, and even a new path might be developed in the process. They also divide technological catching-up into the three conditions of "path creating," "path skipping," and "path following." Wu Xiaobo (1995) thinks that late-developing countries could achieve independent innovation through imitative innovation, creative imitative innovation, and improved innovation on the basis of the study of the dynamic model of secondary innovation. He also thinks that secondary innovation might contain three stages, namely the introduction of technology, its implementation and adoption, and re-innovation; and re-innovation can be further divided into imitative innovation, creative imitative innovation, improved innovation, and late secondary innovation. Wu Xiaobo sketches the evolutionary progress of secondary innovation, as shown in Figure 3-1, and the model and evolution of secondary innovation based on global competition is shown in Figure 3-2.

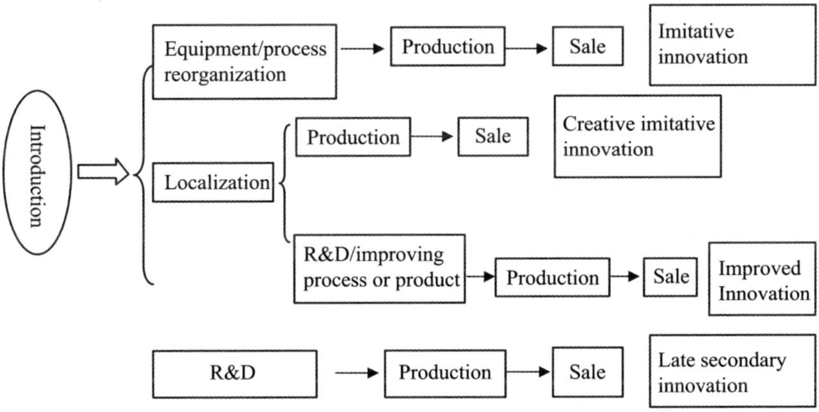

Figure 3-1 The evolutionary progress of secondary innovation

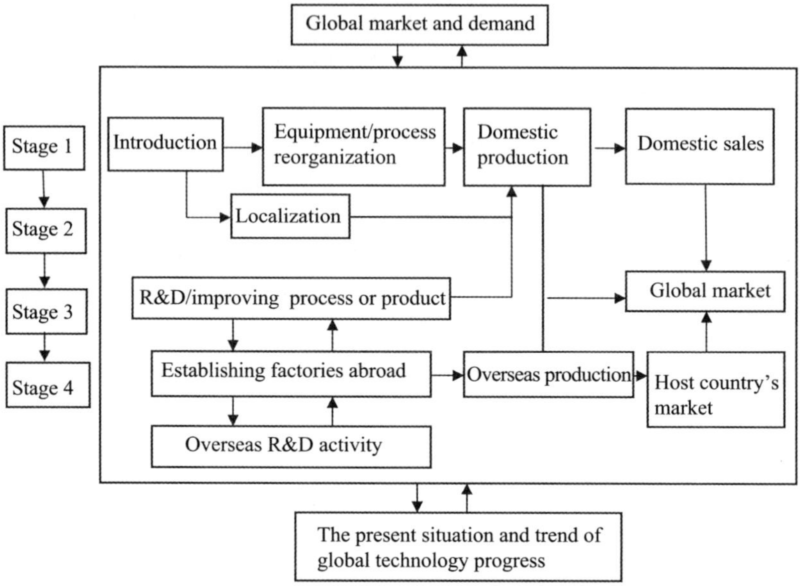

Figure 3-2 The model and evolution of secondary innovation based on global competition

In the former model, the ultimate goal of the introduction of technology is implementation, adoption, and re-innovation. According to experience at home and abroad, if a country's importation of technology depends only on copying, it may get into the cycle of "introduction—falling behind—reintroduction—falling behind again." Only by basing on the implementation and adoption of imported technology, combined with domestic conditions to improve the imported technology and then create new technologies through innovation, can the innovation ability of the country be improved.

Among the foregoing models, imitative innovation is the most basic model of re-introduction. It begins with the introduction of production technology system, including product design, manufacturing process, test methods, material formulas, technical standards, etc. It is based on the principles of imitating the imported technology, which means direct production after a trial run. The enterprise's management strategy is often aimed at establishing advantages by introducing foreign technology to the existing domestic market (Wu et al., 2007). Enterprises at this stage tend to give priority to pure imitation, not really mastering the core technology, but having understood the process of technological industrialization and accumulated knowledge. Accordingly, after analyzing the cost and the consumption of time of innovation and imitation of 48 products in the fields of American chemicals, pharmacy, electronics, and machines, the British scholar Mansfield finds that the average cost of imitation in American enterprises is 65% of the cost of the original innovation, the average amount of time consumed is 70% of that of the original innovation; while the average cost of Japanese enterprises using imitation innovation is 50% of the cost of the original innovation, while consuming 72% of the time. Because imitative innovation reduces the cost and time consumption, it has become a necessary stage for most Chinese enterprises to achieve technological catching-up.

Improved innovation is the fundamental meaning of secondary innovation. In this case, the countries introducing technology develop vertically with the goal of mastering design technology. At the same time, on the basis of technical accumulation, the enterprises begin to form their own technical ability. And then according to the market, the enterprises carry on the improved innovation through independent R&D. That means the "functional understanding" of the imported technology, and then amplifying the application fields of the imported technology, and sufficiently using its "technology opportunity" to develop the new function of the imported products (Xu, 2000). Through improved innovation, the enterprise may get rid of its dependence on foreign technology. However, it is unrealistic to catch up with foreign technology by relying on improved innovation alone. Only by a breakthrough in the original technology paradigm and trajectory and developing new core technology, can countries that import technologies have the chance to realize technological catching-up.

The late secondary innovation is the transitive stage of secondary innovation. The technology introduction of late secondary innovation starts mostly with the introduction of foreign laboratory technology and an involvement in the process of first innovation dominated by the countries that introduced technology through further independent R&D. But this type of innovation requires them to have a higher level of R&D capabilities.

Since the 1980s, China has directly introduced complete sets of equipment or technology from abroad through technological trade. However, with the gradual improvement of the Chinese enterprises' ability for technological innovation, technical cooperation with foreign companies becomes an important path towards secondary innovation. Overall, the model of Chinese independent innovation is still dominated by introduction, implementation, adoption, and re-innovation. The re-innovation centers on imitative innovation and creative imitative innovation, with less amount of improved innovation, especially the late secondary innovation. The Ministry of Science and Technology investigated 2,300 introduced technologies, and found that only 9.2% of the projects that have been introduced have obtained a different degree of implementation and adoption; among the 220 enterprises that were investigated, only 2% of them carried out technological implementation and adoption with scientific research institutes. In recent years, Chinese enterprises have gradually paid more attention to the implementation and adoption of the technology that has been introduced. The introduction of foreign technology cost 39.74 billion yuan in 2004, the technological implementation and adoption 6.12 billion yuan. In 2009, the introduction cost was 42.22 billion yuan, the implementation and adoption 18.21 billion yuan. But compared to Korean implementation and adoption, China still has a long way to go. Among the Korean enterprises that have introduced technology, the proportion of those with entire and nearly entire adoption and innovation is 54.4%; those with basic implementation and adoption account for 36%. Objectively, it is not acceptable to emphasize introduction and have a low opinion of implementation, adoption, and re-innovation. Only on the basis of mastering the imported technology to create new technologies, can China really improve its ability for independent innovation.

3.2.3 Gradual catching-up based on core components (taking Chery as an example)

Among the cases in this book, Chery is a typical case that adopts an innovation model from low-end to high-end.

For a local state-owned enterprise like Chery, it is unlikely to obtain many industrial privileges from the government; the road towards independent innovation is the only choice. Chery found at the beginning of its innovation that the system and the ability for innovation are the resources which accumulate more resources. In accordance with the conventions of industry, if auto enterprises do not produce more than 400,000 vehicles per year, with one billion yuan's fund, it is difficult for them to conduct independent R&D. However, when Chery was "size zero," it had put independent R&D in a primary position of importance. In three crude thatched cottages, Chery successfully developed its first automobile model "Fulwin" with borrowed computers in only two years. When the Fulwin, QQ, Eastar models came onto the market one after another and brought considerable profits, Chery had keenly observed that a group of R&D institutions with core technology had become independent from large foreign auto companies, whose technology spillover provided Chinese enterprises with the chance to realize leapfrogging development. Therefore, Chery determined to fight again. Chery invested from 5 billion to 8 billion yuan in collaboration with the world's top professional companies for the development of new models and key parts.

On December 8, 2001, Chery's 30,000th car rolled off the production line. It was indeed a miracle for this enterprise which had not been producing cars for very long. The Chery Fulwin can be said to be a big hit to the market. The prototype of this car is the Spanish Toledo equipped with a 1.6-liter engine, which belongs to the same class as the Santana, Jetta, and Fukang. But its price was cheaper by nearly a third, and once it appeared on the market, it became very popular. In 2001, Chery sold 28,000 vehicles with 2 billion yuan of sales revenue.

In March 2005, the first high-performance engine was successfully produced, which realized the breakthrough in Chery's R&D and manufacturing of core components and a breakthrough in China. As early as 2002, Chery had already planned and launched the R&D project for 3 series and 18 all-aluminum engines, including the 3-cylinder engine and the V shape 8-cylinder engine, the gasoline engine, and the diesel fuel engine which meets the European IV emission standards from 0.8 to 4.0 liters displacement. High starting point, high standard, and R&D that looks toward the future, all contribute not only to meeting the needs of the domestic market, but also take into account the entrance of Chery's products onto the European and American markets. It will end the situation of consecutive repairs of engines for decades in China, and will become a breakthrough.

At the beginning of 2006, Chery made great adjustments to the models of Fulwin and the new Cowin, and combined them into the new series of Cowin. The high-end models of new Cowin would continue to be equipped with the Tritec engine of the old Cowin model, and the low-end models would be equipped with the SQR480 engine of the Fulwin model. On January 10, the first model of the A520 series, A5, came onto the market. Chery began to participate in the class A family-car market. Up to then, Chery had owned five series models spanning the range from the mini car to the urban leisure SUV. Therefore, Chery had a relatively complete product structure. On March 22, 2006, Chery's first SUV, Tiggo, officially came onto the market. This model is a descendent of a car with an SUV appearance. When it came onto the market, it filled the domestic gap immediately. Because of its excellent cost performance, Chery became the top three on the market.

On June 1, 2006, the sales of the model of QQ, always more than ten thousand cars, was improved with the arrival of the new QQ, equipped with a new 1.1-liter CAC472 engine, on the market, while the Dong'an engine quit the Aqua series. At this point, the Aqua series engine had fully achieved the expectations for it in its independent R&D stage. Moreover, Aqua's cost performance and market competitiveness were greatly improved. In 2006, Chery invested 1.2 billion yuan in new product development and the construction of R&D facilities. The three new models, A5, V5, and QQ6, were put onto the market. The three new models, A18, S12, and S22 also came onto the market one after the other. At the same time, with the support of the National Development and Reform Commission, Chery began to construct engineering laboratories for power transmission assembly and passive security, and sped up the development of hybrid power and the fuel cell, which are the core technology of environment-friendly cars. Therefore, the comprehensive innovation ability was improved further.

In order to break the monopoly of transnational companies in the field of key parts, Chery regarded the cultivation of an independent innovation system for the construction of parts as the breakthrough point. On the basis of introducing international technological talents, Chery invested in the participation and control of some enterprises that produces key parts. Thus, Chery initially developed the ability for independent design, production, and marketing of key components. Smart skylights were developed independently to fill the domestic void. Several technologies of it have reached the international level and reduced the cost of the whole vehicle. All this has laid a solid foundation for the development of Chery's own-brand cars. For a long time, Chery has insisted on taking the road towards independent innovation and independent control of core technologies. Based on independent innovation, the millionth car rolled off the production line in 2007, which proved that independent core technology based on independent innovation can enhance an enterprise's self-restoration ability and is beneficial to the healthy sustainable development of auto enterprises.

At present, Chery has a large number of core technologies, proprietary IPR, and the ability for independent R&D of the whole vehicle, its engine, and some of its key parts. The self-restoration function and core competitiveness have laid a foundation for the subsequent development of Chery.

3.3 The path towards independent innovation based on knowledge-integrated innovation (integrated innovation)

3.3.1 The connotation of integrated innovation

Integrated innovation means that in order to achieve a certain goal, the subject of innovation creatively and organically integrates the related technical elements to explore a new innovative way to create a new product, technology, and process, and even to improve the whole systematical function. The matching of general knowledge with systematic knowledge is the prerequisite for and the key logic around successful integrated innovation. Enterprises need rapid adaptation or reaction speed, so enterprises cannot only consider some single patterns, but also the matching problems of the internal and external technology, namely achieving faster market reaction by inside and outside cooperation. Integrated innovation can be divided into three types—horizontal integration, vertical integration, and network integration. Integrated innovation can help enterprises to deal with discontinuous changes in the market. Some scholars also enrich the connotation of integrated innovation from the perspective of product development management. For example, Pisano (1994) emphasizes the influence of the integration of different sources of technical knowledge within the enterprise on the performance of product innovation. Petroni (1998) also suggests that the concept of "integrating meta-competence," which expands the Strategic Management Theory based on dynamic capabilities. Hardaker (1998) decomposes integrated innovation into integration at a technological level, at a managerial level and at a cultural level, after having studied the product development and management of European manufacturing enterprises. For the first time, he introduces management into integrated innovation as an element of integration, and he thinks that integrated innovation is the result of the integration of multiple departments, functions, and organizations.

Chinese scholars basically inherit such point of view on the understanding of the concept of integrated innovation, which means not just focusing on innovation through technological integration, but also on some elements, such as management. The first is the study of the integrated innovation model. Li Baoshan & Liu Zhiwei (1998) hold that integration refers to a kind of creative fusion process from the point of management, namely focusing on creative thought in the process of the combination of all elements. Chen Jin (2002) claims that integrated innovation is the sum of technological integration, knowledge integration, and organizational integration. He also puts forward a new framework that analyzes integrated innovation, namely the analysis framework of strategic integration, knowledge integration, and organizational integration. Cao Lijun et al. (2005) affirm that integrated innovation is a systematical and dynamic concept. The performance of the overall system function depends on the structural proportion and the coupling mechanism of the elements.

3.3.2 The inner structure of integrated innovation

With the growing complexity and the increasing competition in the field of technology, the traditional innovation model has been unable to solve innovation problems of a complex nature. More technologies and knowledge must be integrated into the same innovative architecture. But integrated innovation is not only a simple integration of multiple technologies and knowledge, because it involves the dynamic adjustment of technology, knowledge, and organization both inside and outside the enterprise, when technological integration is carried out. The latter could be called organizational integration.

(1) Technological integration

Technological integration often involves the integration of self-owned technology, technology that has been introduced from outside, and technological learning. It is actually an effective measure and method an enterprise employs that takes its own core technology as the supporting element, the market demand as a guide, and introduces other advanced technologies or learns by consulting technical data, so as to highly integrate all kinds of technologies into one product. Consequently, an enterprise realizes the development of new products in a short time, arrives first and as rapidly as possible on the market, and then acquires a large share of the market in a short time. By improving the model of Jiang Hui & Chen Jin (2000), we can get the model of technological integration of integrated innovation as shown in Figure 3-3.

Technological integration is the foundation for and also the starting point of integrated innovation. Thus, on the one hand, we need to notice whether the integrated technology can meet the requirements of product innovation or process innovation; on the other hand, we should also make it as clear as we possibly can that what the inherent requirements are for enterprise knowledge integration and organizational integration.

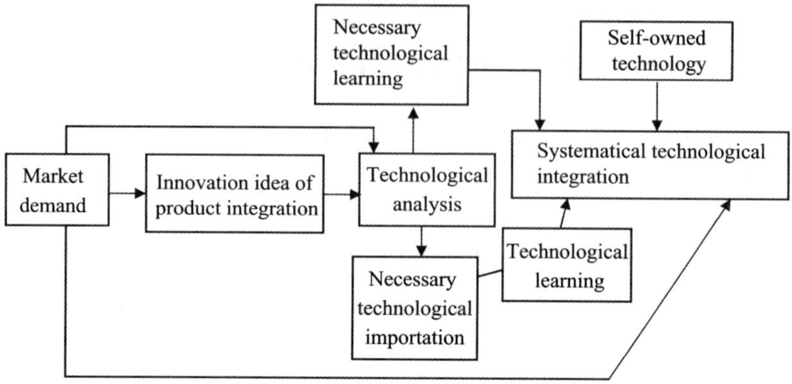

Figure 3-3 Technological integration of integrated innovation

(2) Knowledge integration

Knowledge integration is a must in integrated innovation. Because of the progress of product or process innovation by using technologies of multiple sources, personnel of different levels in the enterprise are faced with the problem of knowledge integration. The first person proposes the idea of knowledge integration is Professor Ikujiro Nonaka (1991). He claims that enterprise knowledge integration could be regarded as the interaction between tacit and explicit knowledge, which suggests four basic patterns, namely the transformation from tacit to tacit, from tacit to explicit, from explicit to explicit, and from explicit to tacit. Tacit and explicit knowledge run on two dimensions of epistemology and ontology, and they move constantly and interact with each other. This process makes knowledge flow from tacit to explicit and finally return to tacit, which forms a spiral escalation of knowledge through innovation. Knowledge integration is seen as the integration of knowledge, namely the dynamic integration of explicit and tacit knowledge in different stages in the process of enterprise innovation, and the integration of knowledge subjects, namely the integration among the different subjects on the same or different levels. Within the enterprise, knowledge integration mainly presents the process through which the individual knowledge becomes group knowledge through communication and then gradually a part of enterprise knowledge. In integrated innovation, knowledge integration beyond enterprise boundaries is necessary; this refers to the intersection of an enterprise's internal and external knowledge. Obviously, once the integrated technology includes external technology, enterprises need to acquire external knowledge associated with the technology.

(3) Organizational integration

In integrated innovation, technological integration puts new requirements on knowledge integration and organizational integration. Organizational integration is a process in which the enterprise integrates its organizational elements with different functions (unit organization or independent organization) as an organic organization with relatively complete functions for the enterprise to maximize the implementation of integrated innovation. Compared with the traditional linear innovation organizations (Figure 3-4 a), the organizations of integrated innovation are similar to a matrix (Figure 3-4 b). In linear innovation organizations (Figure 3-4 a), the enterprise's research department focuses on creating and selecting technologies, then gives these to the technological development department in order to achieve the productization or technicalization of the developed technology. In this case, the research department and the development department must in turn make a great effort independently in order to make it possible to achieve innovation to some extent. In integrated innovation organizations (Figure 3-4 b), the research department takes charge of creating and acquiring technology with the cooperation of other departments; the research and development departments take charge of assessing and introducing technology with the cooperation of other departments; the development departments take charge of improving the technology that has been created and introduced with the cooperation of other departments; the development and production departments are in charge of the productization of the technology that has been created and introduced with the cooperation of the marketing department. Organizational integration involves a process of close coordination among organizations. It is possible for the enterprise to realize the desired innovation efficiently.

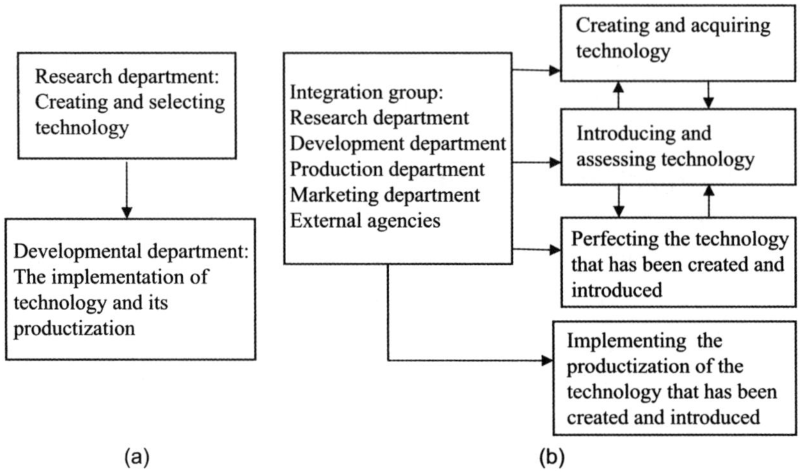

Figure 3-4 Comparison of the traditional innovation organization (a) and integrated innovation organization (b)

3.3.3 Integrated innovation based on end products (taking FAW Car as an example)

The achievements of integrated innovation may not only be product innovation, but also process innovation. Taking product innovation as the background and taking FAW Car as an example, we will discuss integrated innovation based on end products.

FAW Car Co., Ltd. (FAW Car for short) was founded in 1997 and has two independent brands—Hongqi and Besturn, based on the inheritance of the technological accumulation of the FAW Group. In 1958, FAW Car produced the first Chinese luxury sedan Dongfeng CA71 by imitating the 1955-type Chrysler. Then after five systematic experiments, CA72 came out. Its name was changed to Hongqi (commonly known as "Big Red Flag"). During the next 20 years, FAW Car developed CA770, CA771, CA772, CA773, CA770W, and CA770 one after another. Then the "Big Red Flag" series stopped production.

At this stage, the innovation model of FAW Car was "independent development based on imitation." For example, regarding the model CA772, FAW Car developed the engine, gearbox, drive axle, and the whole transmission system and chassis, and achieved knowledge accumulation of the architectural style of the whole vehicle and its core components knowledge. However, there was still a huge gap between FAW Car and the advanced level of the world. FAW Car was at the stage of manufacturing cars in a completely manual way.

To promote the level of technology rapidly, FAW imported the complete set of equipment for the 488-type engine and all the technical drawings in 1987 (there were no technical know-how), and built the engine production line in a real sense without implementation or adoption on a large scale. 488-type engine was reformed and equipped into the later Hongqi CA7220 (commonly known as the "Small Red Flag"). In 1988, in cooperation with the German Volkswagen, FAW Car introduced the vehicle assembly technology of the Audi 100 and achieved short-run production. This promoted the learning of architectural knowledge, but core components such as the engine were still controlled by the German side. FAW Car matched the Audi 100 body with the 488-type engine and the 016-type gearbox as well as a self-developed cooling system, exhaust system, electrical system, gear train, stents, and engine hood in 1996, and then produced the new Hongqi CA7220. In 1998, FAW Car matched the 4GE Engine with the "Small Red Flag," and then launched the Hongqi 98 Star. In 2000, FAW Car matched the Audi 200 chassis with the Nissan 92 kW 2.0L V6 engine, and then launched the new product—Hongqi Century Star. The integrated innovation of a series of vehicles strengthens the architectural knowledge of FAW Car.

With the development of the Chinese car market, the Audi product platform of the 1970s was unable to meet the market demand. Therefore, in 2002, FAW Car introduced the assembly line and management system of the Japanese Mazda M6 by means of technology licensing, followed by the vehicle assembly. All of this brought about the transition to mass production. But the core components of the M6, such as the engine and the chassis, were still controlled by the Japanese side. In order to make full use of the M6 product platform and to develop the ability for integration at the product system level, FAW Car decided to depend on the product platform with M6 core components to develop a new product with the working code C301 in 2002. Based on the model design of the Italian company IDG and integrated with the M6 core components, such as the engine and the chassis, in 2006 FAW Car officially launched the independent brand car Besturn, realizing integrated innovation at the whole vehicle level once again. In the process of product development, in addition to being responsible for the matching of non-core components to core components, FAW trained the technical personnel of the Japanese Mazda to complete the matching of the engine, the chassis, the gearbox, and the clutch, which accumulated knowledge architecture for the further use of the M6 product platform in integrated innovation. In 2006, the improvement project of Besturn with the code C302 started; it was expected to carry on a further development of the chassis, the powertrain, and the electrical system and realize a great leap ahead in knowledge accumulation of the core components.

FAW Car fully realized that it is very important to master framework technology for the development of new products before fully mastering the core component technology; especially when vehicle matching has become an important direction in development. The first domestic gasoline direct injection engine, the JB8, developed by the FAW Group's R&D center, succeeded in 2006. In 2007, the gasoline direct injection engine GDI was successfully developed; and based on this, FAW Group researched and developed the high-end car engine independently. In addition, according to the national 863 platform, the Group technology center researched and developed the Hongqi car with hybrid power through the cooperation and joint development of industry, university, and research, in order to promote the industrialization of electric cars.

In its development process, in order to narrow the technological gap with foreign companies as soon as possible, FAW Car gave up the early strategy of "independent development based on imitation" and turned to the acquisition of mass external technology. But too much attention to the market led FAW Car to only focus on the terminal link of vehicle technique. Although it has launched a series of products of the independent brands like Besturn and Hongqi and occupied some market share through integrated innovation, it has not fully mastered the core component technology, which has caused dependence on an external vehicle product platform and continuous introduction of assembly technology, which means the "non-core" situation of technological development. Although FAW Car is using the technology resources of the Group to gradually build an industrial chain technique, there is still no fundamental breakthrough of the industrialization of gasoline engine technology and hybrid power car technology. Therefore, the development of FAW Car technology is still an integrated innovation of end product.

Throughout the technological development of the car enterprises with independent brands, the catching-up of technological innovation based on the construction of a technological chain has appeared. Since then, the Chinese company Chery has begun to imitate and develop the Fulwin, based on the integration of the Ford CVH engine. Depending on cooperative development and imitation strategy, Chery has successfully developed, in an integrated manner, the Eastar (equipped with a Mitsubishi engine), the QQ (integrated with a Dong'an engine), the Cowin (equipped with a Chrysler engine), and the Tiggo (integrated with a Mitsubishi engine), and the company has hence mastered the technology of product architecture. Then, in addition, through cooperation with the Austrian company AVL, Chery developed 18 ACTECO series engines to improve the present car models. At the same time, the Chang'an Company is also a typical example of integrated innovation in China. Based on the model design of the Italian company DEA and the integration of a Ford engine, Chang'an successfully developed the CM8 model. After cooperation with the German engine technology company FEV, Chang'an developed the 2.0L engine successfully, which was equipped for the newly developed CV6, CV8, and CV9 models.

3.4 The path towards independent innovation: Innovation (original innovation) directly from high-end

3.4.1 The connotation of original innovation

The core characteristic of original innovation is having independent intellectual property and an independent brand. Thus, it is the most competitive independent innovation model, the most fundamental and the most productive kind of innovation.

Xu Guanhua (2001), the former Minister of the Ministry of Science and Technology holds the opinion that original innovation means the unprecedented invention or discovery in the area of R&D, especially in the fields of fundamental research and high-tech research. So its innovation achievements not only prolong the innovation period, but also create new innovation periods and an innovation climax. Wu Haijiang (2002) thinks that original innovation means making unprecedented contribution to the scientific community. The application value of original innovation achievements might hardly be noticed at the time, but it can lay a foundation for future scientific development. Chen Yalan et al. (2003) believe that original innovation is a process, in which the phenomenon, structure, movement, and the law of interaction are explored by scientific experiments and theoretical research, or critical scientific problems in the development of the economy and society are solved by scientific theories. Its achievements are significant scientific discoveries, great theoretical innovations, significant technological innovations, and inventions of experimental methods and instruments. The experts of the National 973 Program (2004) also give a definition of original innovation from the two fields of science and technology. They claim that original innovation in science means exploring new fields, leading in new directions, and cultivating new disciplines. Its achievements include the discoveries of new phenomena, and the formation of new concepts and theories. While original innovation in technology means great breakthroughs in technology. Its achievements include great inventions of new methods, process, and products. They think that original innovation in science would have an epochal influence on technological development and human society as well; original innovation in technology often leads to great changes in industrial fields and also promotes original innovation in science itself. Chen Jin & Xie Lianghong (2004) hold that original innovation is the result of long periods of accumulation, that it has great possibilities for exploration, albeit with much uncertainty, it is the pioneer of new technology and inventions.

According to these scholars' definitions of original innovation, we can summarize its characteristics as follows: The first one is independence, which emphasizes the initial and voluntary actions of the subjects of innovation. The independence of the original innovation does not mean the process of original innovation is completed by a single subject of innovation. It still emphasizes communication and cooperation among different subjects. In general, original innovation derives from the fields of fundamental research or applied fundamental research, positions itself at the forefront of science, and has a high requirements for the quality of researchers. It requires communication and cooperation from different fields. In the process of original innovation, it is beneficial to mobilize the initiative of different subjects of innovation through technological alliances such as the industry-university-research alliance. It can also achieve a sharing of resources and advantages, which may reduce and dispel the risk of original innovation, and then promote its rate of success. The second characteristic is originality. Original innovation must first propose concepts, theories, discoveries, and inventions, which has great promise for scientific development. The third characteristic is breakthrough. Original innovation should be breakthrough innovation, which means a great breakthrough from traditional theories, opinions, and technologies and a role played in promoting social development. The countries which realize more original innovations are more competitive in the fields of S&T, and also have a leading position in the trend towards world development. Moreover, it is easier for enterprises which make original innovations to obtain advantageous positions in market competition.

3.4.2 The realization path of original innovation

G. Wallas (1926) claims that all inventions and creations have four stages—preparation, inoculation, illumination, and verification. The preparation stage includes finding problems, collecting information, arranging and analyzing information, and understanding the achievements of predecessors as well as some unsolved problems, so as to lay the foundation for the next innovation. The inoculation stage means leaving the former problem unfinished and thinking about different problems from different perspectives in order to find different possible solutions. The illumination stage is that creators will have a better and clearer understanding of the problems they had thought about and have an instantaneous "flash" after a period of inoculation. This stage is the result of full preparation and a long period of inoculation of the two preceding stages. The verification stage is to test the validity of the inspiration and idea. If the new perspective does not tally with the assumed result, the innovation has to start over again. W. I. Beveridge (1979) puts forward the 4C theories, namely collecting, contemplating, conceptualizing, and criticizing. Wallas' and Beveridge's four-stage-theory have many similarities as can be seen from the above statement.

The essence of original innovation is creation. In the process of original innovation, the accumulation of existing knowledge is the source. It becomes the motivation factor for original innovation together with project selection. Core talents, the innovation culture, the incentive mechanism, the original technique, research interest, and teamwork are important factors which will influence original innovation. If original innovation succeeds, the protection of intellectual property will be very important. Of course, if it fails, the reason for the failure should be found and a new endeavor should begin.

Original innovation is based on knowledge production. Its occurrence mechanism is shown in Figure 3-5. A typical original innovation starts with fundamental research, including pure fundamental research and applied fundamental research. Pure fundamental research evolves into the original innovation of it through the spiral ascension (P-OI); applied fundamental research is based on the knowledge-converting process of "metaphor—analogy—model," and ultimately evolves into the original innovation of applied fundamental research (A-OI); both of them interact and influence each other through the production and spread of knowledge. The P-OI and A-OI also interact during the process (Shu Chengli & Gao Shanxing, 2008).

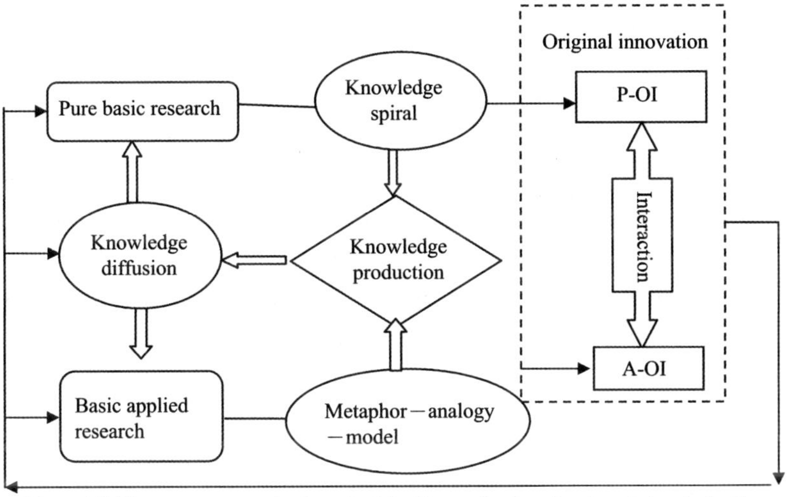

Figure 3-5 The occurrence mechanism of original innovation based on knowledge production

On the basis of "the occurrence mechanism based on knowledge production" (shown in Figure 3-5), we can conceive a more detailed model—"the realization mechanism of original innovation" (shown in Figure 3-6). The occurrence and realization of original innovation can be generalized as follows: First, on the basis of the accumulation of existing knowledge, innovators choose projects and invest resources according to the changes of the domestic and international environments and the requirements for competition; second, according to the former choice and investment of resources, innovators carry out corresponding pure fundamental research or applied fundamental research, and then get some original innovation achievements through the process of knowledge spiral or analogy—metaphor; third, once the achievements of original innovation are accepted, innovators need to depend on the intellectual property system to protect the fruits. If they are not accepted, a new round of innovation should be carried out; fourth, the accepted original innovation achievements can be used as the knowledge foundation for the next original innovation.

Since the reform and opening up, the Chinese government and

enterprises have been gradually paying more attention to original innovation with the development of the economy. However, compared with developed countries, there is little original innovation in China. There might be some reasons: The first is the weak foundation and low starting point; second, under the policy of "exchange for market," the motivation and investment in performing original innovation is not sufficient. Taking Chinese fundamental research fund expenditure as the case, we can easily see the difficulties of performing original innovation. Especially, although the fundamental research expenditure of R&D increased from 1.81 billion yuan in 1995 to 5.22 billion yuan in 2001, and 26.48 billion yuan in 2009, which is 14.63 times of the expenditure in 1995, the proportion of fundamental research expenditure in R&D expenditure has a decreasing tendency from 5.19% in 1995 to 4.75% in 2009, a decrease of about 0.62%.

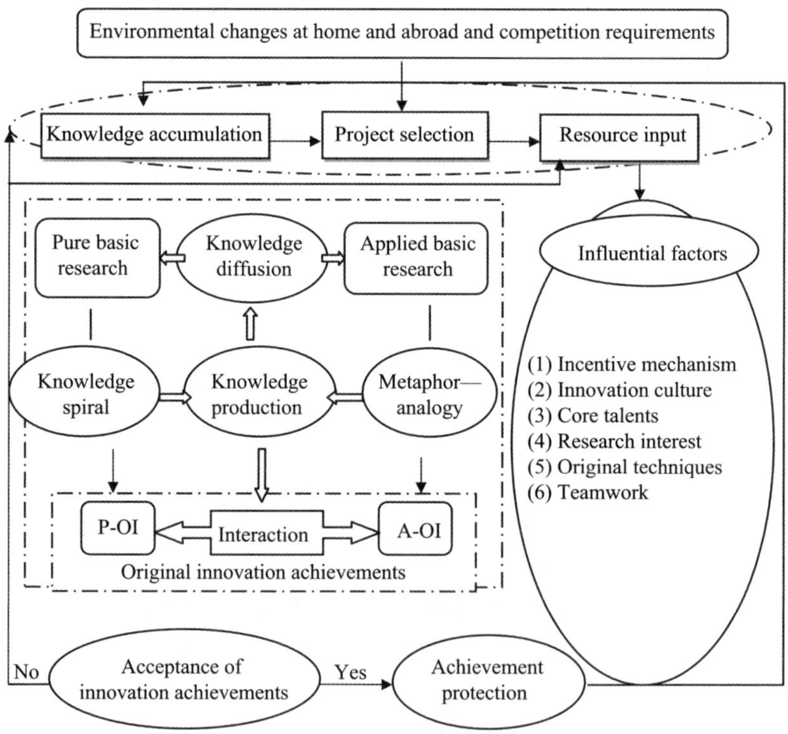

Figure 3-6 The realization mechanism model of original innovation

3.4.3 Market-oriented innovation based on technological breakthrough (taking Suntech as an example)

For a long time, the Chinese utilization of solar energy has remained at the initial stage dominated by solar energy water heaters. In recent years, with the implementation of the developmental strategy of non-fossil fuel energy, some domestic enterprises and entrepreneurs have started to promote the development and production of solar energy generation equipment. The founder of Suntech, Dr. Shi Zhengrong, studied in Australia at the University of New South Wales in 1998, and studied with the international solar energy cell authority and the Nobel Prize laureate, Professor Martin Green. In the period of studying abroad, Shi won more than 10 technology invention patents regarding research on the polycrystalline silicon solar thin film battery. He got a PhD in 1991 by working on the technology for the polycrystalline silicon solar thin film battery. In 1995, he held the post of company director of Australian Pacific Solar Power Co., Ltd. owned by Professor Green. In 2000, Shi went back to establish Wuxi Suntech Power Co., Ltd. (Suntech in short). Suntech launched its IPO on the New York Stock Exchange on December 14, 2005.

Suntech has many regional branches in the world, located in China, Japan, Australia, UK, Germany, South Africa, the US, etc., and five major manufacturing bases in Wuxi, Shanghai, Luoyang, and Xining in China, and Nagano in Japan. And now it is one of the leading global solar photovoltaic (PV) enterprises which specializes in product R&D, manufacturing, and marketing of the transistor silicon solar energy battery and its components, the silicon thin film solar cells, PV power systems, and the building of integrated photovoltaics (BIPV). The innovation of Suntech is a typical example of “market-oriented innovation based on technology breakthrough.” The breakthrough in technology includes about a dozen technology invention patents earned by Shi Zhengrong in the period of his studies at NSW University; "market orientation" is the strong demand for developing solar energy in China.

Suntech specializes in the PV industry that turns solar energy into power through the solar energy battery. Its industrial chain is very long. The upper stream is crystalline silicon manufacturing; the intermediate stream the solar battery pieces which will be assembled into the solar cell component after a complete series of packaging processes; the lower stream the solar energy equipment manufacturers. Suntech has chosen to specialize in the research, manufacture, and sale of the solar energy cell, its components, and the PV power system. Oriented by the market, Shi and his Suntech choose to produce traditional crystalline silicon batteries, combining with Chinese manufacturing advantages.

In order to win the market competition advantages sustainably, while establishing an R&D team, Suntech also has a long term cooperative R&D relationship with the PV center of the Australian NSW University and Zhongshan University. Thus, Suntech has independently developed the crystalline solar energy cell with high efficiency and low cost, as well as PV components, etc., a total of 20 core technologies with independent intellectual properties. Besides the independent innovation of crystalline silicon solar energy cells and PV components, the independent innovation of Suntech is also shown in transforming foreign equipment. While adapting to the requirements of the complex process, the cost of the second and the third battery production line is 40% and 50% less compared to the first line, respectively. Based on this, the rate of transformation of the polycrystalline silicon solar cell reaches 17.3% and 15.5% respectively. In the same scale of production, the rate of foreign competitors is 16% and 14.5% respectively. Suntech is ahead of its foreign counterparts.

Based on the innovation stated above, Suntech caters for the needs of both the domestic and the foreign markets by virtue of its solar PV products. In August 2002, the first solar production line with 10 MW was put into production, its manufacturing capacity reached 60 MW in 2004, putting it into the list of the world's top 10 solar cell manufacturers, its total output ranking eighth in the world. Its sales revenue increased from $3 million in 2002 to $85.3 million in 2004, and compound annual growth rate was as high as 430.9% from 2002 to 2004. In 2009, of Suntech's component production output of the crystal silicon solar cell reached 1,100 MW, and its annual export volume reached 704 MW, thus becoming one of the world's largest manufacturers of crystalline silicon solar cell and components. Some people in this industry think that Suntech's rise has shortened the gap by 15 years between China and the world in the solar PV industry.

References

Beveridge, W.I.B. *Seeds of Discovery: A Sequel to the Art of Scientific Investigation*. Portsmouth, NH: Heinemann Educational Publishers, 1979.

Cao, L. Research on enterprise independent innovation system and model. Doctoral dissertation of Hefei University of Technology, 2006.

Cao, L. et al. Trajectory analysis of high-tech transformation of traditional industry: The concept, mode, and system of integrated innovation. *Science & Technology Progress and Policy*, 2005, 22(5): 51-53.

Chen, J. & Xie, L. Research review of original innovation. *Science of Science and Management of S.&T.*, 2004 (2): 23-26.

Chen, Y. et al. The research on the main factors and the evolving mechanism of original innovation. *Studies in Science of Science*, 2003, 21(4): 433-437.

Hardaker, G. An integrated approach towards product innovation in international manufacturing organizations. *European Journal of Innovation Management*, 1998, 1 (2): 67-73.

Jiang, H. & Chen, J. Integration innovation: A new innovation mode. *Science Research management*, 2000, 21 (5): 31-39.

Kim, L. Imitation to innovation: The dynamics of Korea's technological learning. *Journal of International Business Studies*, 1997, 28 (4): 868-872.

Lee, K. & Lim, C. Technological regimes, catching-up and leapfrogging: Findings from the Korean industries. *Research Policy*, 2001, 30 (3): 459-483.

Li, B. & Liu, Z. *Integrated Management: The Management Innovation of High-tech Era*. Beijing: China Renmin University Press, 1998.

National 973 Program of Basic Research in Common Major Issues Strategy Team. Some problems about performance evaluation of basic research. *China Basic Science*, 2004, 6 (3): 3-8.

Nonaka, I. The knowledge-creating company. *Harvard Business Review*, 1991, 69 (6): 96-104.

Petroni, A. The analysis of dynamic capabilities in a competence-oriented organization. *Technovation*, 1998, 18 (3): 179-189.

Pisano, G.P. Knowledge, integration, and the locus of learning: An empirical analysis of process development. *Strategic Management Journal*, 1994, 15 (S1): 85-100.

Shu, C. & Gao, S. The generating mechanism of original innovation: From the knowledge creation perspective. *Studies in Science of Science*, 2008, 26 (3): 640-644.

Sun, B. & Shang, Q. Technological innovation pattern suitable for our enterprises—Secondary innovation. *Scientific Management Research*, 2002, 21 (2): 20-22.

Wu, H. Original innovation in basic research and the accumulation in science. *Studies in Dialectics of Nature*, 2002, 18 (5): 42-50.

Wu, X. The evolution of the secondary innovation process. *Science Research Management*, 1995, 16 (2): 31-37.

Wu, X. & Ni, Y. Secondary innovation and strategies for China's manufacturing industry in the global competition. *Science Research Management*, 2001, 22 (3): 41-45.

Xu, G. Emphasizing on basic research and promoting original innovation. *China Awards for Science and Technology*, 2001, 9 (2): 1-6.

Chapter 4

The capacity, mode, and trajectory of Chinese independent innovation

4.1 The connotations of the capacity, mode, and trajectory of innovation

4.1.1 Innovation and independent innovation capacity

(1) Innovation capacity

Innovators must have the appropriate capacity to innovate in order to carry out innovation. The innovation capacity is the ability to effectively adopt and control the knowledge and skills needed for innovation, the improvement of existing technologies, and the creation of new technologies (Lall, 1992). In recent years, academic research on innovation capacity has been understood and expanded at three levels, that is, nation, industry, and enterprise.

Among them, the innovation capacity at the national level is a country's ability to create and commercialize the world's leading technology in the long run. This ability is determined by a common innovation infrastructure and specific technology and technology cluster, as well as the quality of the links between them (Furman et al., 2002). At the same time, the MIT journal *Technology Review* published an article in which it is affirmed that the capacity for national innovation is "the degree to which a country provides a favorable environment for promoting world-wide cutting-edge technology," which represents the potential of a country to produce innovation achievements with a commercial value. In the report "Global Innovation/National Competitiveness," the US Center for Strategic and International Studies (CSIS) points out that the connotation of the capacity for national innovation far exceeds the strength of investment in S&T. It is the synthesis of many successful elements, including a great investment in S&T; the knowledge, skills, and talents who master and use them; an investment-friendly environment for innovation; powerful industrial investment in new technologies and products; well-educated workforce and professional scientists and engineers; networks of cooperation that connect users to R&D institutions; an open trading system. In short, it is not difficult to see that the essence of the capacity for national innovation is the ability to participate in international innovation competition through institutional, technological, and management innovation, which are carried out by the innovation subjects, including governments, enterprises, universities, research institutes, financial institutions and agents, etc.

As for industrial innovation and the capacity to carry it out, Shi Qingqi & Shang Yong (2002) claim that industrial technological innovation refers to the sum of a series of activities, which are market-oriented and take the improvement of industrial competitiveness as the goal for the new generation of products or process via the acquisition of technology (research, development, technology introduction, implementation, and adoption), construction, and industrialization. Accordingly, the capacity for industrial technological innovation is the ability to promote industrial development through the introduction or development of new technologies.

Accordingly, the so-called capacity for enterprise technological innovation is the synthesis of various abilities supporting the implementation of the innovation of enterprise technology. It includes the ability to make use of the resources and to understand industrial and technological development, related structure, and cultural conditions, and the capability for strategic management (Burgelman et al., 1995). Szeto (2000) believes that the capacity for enterprise innovation refers to the overall ability to develop new products, meet market needs, and implement innovation. Westphal et al. (1981) hold the opinion that the capacity for enterprise innovation is the synthesis of an enterprise's organizational and adaptive abilities as well as its ability to acquire technology and information. Steven Muller claims that the capacity for enterprise innovation is the synthesis of the capability for product development and production technology improvement together with technical reserve and organizational abilities. Burgelman thinks that the capacity for enterprise innovation is made up of available resources, the ability to understand industrial competitors and the environment, the company's organizational culture and structure, and its pioneering strategy. In short, it is obvious that the so-called capacity for enterprise innovation is essentially the organic combination of an enterprise's abilities to implement technological innovation and achieve commercialization.

(2) The capacity for independent innovation

The capacity for independent innovation means a nation's, an industry's, or an enterprise's ability to independently promote and implement the innovation which may form intellectual property. It emphasizes a decision for, and the implementation of innovation and the ownership of property. If China wants to have more innovation now and in the future, it must first improve its capacity for independent innovation at the three levels of nation, industry, and enterprise.

Since China proposed the strategy for the implementation of independent innovation, many scholars have discussed capacity for independent innovation. Mao Yunshi & Wang Jiancheng (2006) claim that independent innovation is more a process of introduction, implementation and adoption, imitation and innovation. It is a process of accumulation and learning, in which innovators can form their own ability for independent innovation. Song Hefa et al. (2006) think that the capacity for independent innovation means the innovation subjects' ability to acquire leading innovation property rights and innovation earnings through active efforts, and then get a long-term competitive advantage. Innovation property, he says, mainly refers to technological IPR like patents. And innovation earnings include innovation profit and technological progress. Guo Jiucheng & Zhu Konglai (2008) believe that the capacity for independent innovation is the result of the interaction of various abilities throughout the whole process of innovation, including the main ability necessary for innovation to master and use resources; the support and contribution of the external environment, which help to realize the ability to grasp and use innovative resources; the creative ability of S&T achievements (productive competence); and the cultivation ability of new products and market brands. The former vice president of the Chinese Academy of Science and Technology for Development (CASTED), Professor Wang Yuan (2005), holds that the ability to improve independent innovation is a process of economic and market implementation.

The foregoing discussions do not give any details about the level of the capacity for independent innovation, so the discussion applies for nation, industry, or enterprise to some extent. Of course, there are scholars who discuss explicitly the capicity for independent innovation. Xia Zhiyong & Yuan Jianhua (2007) point out that the enterprises' capacity for independent innovation means the synthesis of various abilities of acquiring independent IPR, enhancing enterprise core competitiveness and keeping sustainable competitive advantages by making full use of internal and external innovation resources, establishing a new technological platform or changing the core technology. Lu Feng (2006) believes that it is difficult for Chinese enterprises to carry out independent innovation independently under the condition of a significant dual structure. The enterprises also need the support of the government. In other words, governmental support is also an important part of the capacity for independent innovation.

4.1.2 The mode of enterprise innovation and the trajectory of industry innovation

When people talk about mode of innovation, they often refer to the enterprise; and when they talk about trajectory of innovation, they often refer to the industry. This will be discussed below.

(1) The mode of enterprise innovation

A mode refers to systematical ideas and methods for solving problems in general. To summarize, concluding and promoting the methods of solving certain problems forms the so-called mode. And innovative mode or innovation mode means the concrete way and method in which innovators think and implement systematically under specific conditions and circumstances. Based on a feasible innovation mode, enterprises can achieve their original intention of innovation, otherwise it may backfire.

Concerning the mode of innovation, previous researches have discussed linear and non-linear innovation modes. The linear innovation mode means that the enterprise realizes innovation in a unidirectional and progressive process of "applied research→design and trial manufacture→manufacture→sale" based on basic scientific knowledge. However, a lot of innovation practices prove that the linear innovation mode is not consistent with the practical process of innovation because innovation is non-linear. There are complex linkages and feedback among the various stages of innovation. Thus, some scholars have summed up the non-linear innovation mode. However, whether the mode is linear or non-linear is not important for the innovation mode discussed in this book which focuses on the realistic problem regarding what systematical innovation ideas and ways an enterprise should have so as to be able to solve the problems during the process of innovation in the current global competitive environment with the current technological foundation and innovation capacity in China.

(2) From a technological trajectory to a generalized trajectory

The technological trajectory of innovation was first put forward by Dosi in 1982. Compared with the term "scientific paradigm" put forward by the scientific philosopher Kuhn, he proposed the concept of "technological paradigm." He claims that the technological paradigm is "a model or mode for solving technical problems," which determines the field of research, the issues, the procedures, and the tasks. It is not only a conceptual system of examining the problem, but also a system regarding the way of solving the problem; it stipulates the direction and the connotation of technological innovation in a given period. According to technical paradigm, Dosi (1982) defines the technological trajectory as the regular direction, connotation, and method decided by paradigm to solve technical problems. It is a set of possible technical directions; its outer boundary is determined by the nature of the technological paradigm. Dosi (1982) thinks that innovation is always based on a technological trajectory in a specific period of development and in a specific field, which means the evolution of innovative technology is dominated, supported, or restricted by a technological paradigm. Innovations in different industries or enterprises have different technical trajectories at some point. For example, Saviotti finds that the innovation of the airplane had two different technical trajectories from 1915 to 1983—one focused on the large body (rather than on speed), such as the Boeing 747, the McDonnell Douglas I, the Douglas DC-10, etc.; the other one focused on speed (rather than on the body), such as fighter aircrafts or supersonic aircrafts.

With the support of the National Natural Science Foundation, Du Yueping et al. (1995) did early research on the domestic technological trajectory of innovation (1991–1993). Through imitating the "technological trajectory" theory contemplated by Dosi (1982), Du Yueping and Lei Jiasu (1995) propose that innovation should not only solve the problems in the source of technology, but also the problems in the selection of technology, product and process design, technology standards, investment patterns, and the selection of the marketing program, which are related to the application of the new technology; in a certain period and in a specific field, the solution to these problems will be controlled by a certain paradigm. Therefore, objectively speaking, innovation is often controlled by multiple paradigms in various fields. On the basis of Dosi's (1982) point of view from a technological paradigm to a technological trajectory, it can be considered that the majority of enterprise innovation within an industry is actually determined by the generalized trajectory, which is made up of the dominant paradigms of technical source, new technology selection, product design, process design, technology standards, field investment, and marketing. In other words, the so-called generalized industrial innovation trajectory determines whether industrial innovation will develop according to a particular direction, connotation, and method. Coincidentally, according to people's pursuit for product performance and the current situation of technological development, another economist, Biondi, sums up the innovation technological trajectory of most enterprises in various industries, namely that of reducing cost, improving life, developing scale economy, using resources efficiently, realizing market segmentation, and fast service.

4.2 The mode and the building up of the capacity for enterprise independent innovation

Enterprise independent innovation is a process of value creation and realization from R&D via commercialization and construction of new technology to manufacturing and market realization. Only when the enterprise possesses the ability to carry out original independent innovation, can it produce more innovations in order to achieve sustainable development. Based on all the above, more enthusiasm is needed to focus on the formation of the ability, the mechanism, and the mode of enterprise independent innovation.

Combined with the situation of the innovation of Chinese enterprises from 1998 to 2003, Wang Lina et al. (2006) divide innovation into three types based on the source of and the path towards innovation: i) The process of the improvement of innovation capacity of "technological introduction—implementation and adoption—independent innovation" based on imitation innovation; ii) the process of the improvement of innovation capacity of "independent R&D—incremental innovation—fundamental innovation" based on independent innovation; iii) the process of the improvement of innovation capacity of "technical cooperation—implementation and integration—independent innovation" based on cooperative innovation.

4.2.1 The mode and the improvement of capacity in different stages of enterprise innovation

In different innovation stages, the independent innovation mode and the building up of the enterprises' capacity are different. For example, the stage of enterprise innovation can be divided into "improved innovation, penetrated innovation, integrated innovation, and breakthrough innovation." In the different stages, the characteristics of the innovation mode are different (shown in Table 4-1).

Table 4-1 The mode and corresponding capacity for enterprise innovation in different stages

Innovation stage	Innovation capacity	Innovation mode
Improved innovation	Ability to learn and improve	Improving product and market development
Penetrated innovation	Ability to expand the field	Developing multi-type and serialized product
Integrated innovation	Ability to integrate market segments and abilities to segment	Integrating related capability and developing new product
Breakthrough innovation	The capacity for matrix-type innovation and the ability for special innovation	Providing new value to the market

(1) Stage of improved innovation: Improving the product and developing the market

On the basis of the formation of a certain capacity for technological innovation, the enterprise can carry out improved innovation, namely innovation dominated by product and market development. In this stage, the enterprise must find and improve the imperfections of the product based on customer demand. Then, based on investigating the concept of the new improved products, enterprises develop an improved product and make an effort to develop the market. So-called market development means that enterprises develop new products, look for and lock in target customers through a series of marketing strategy, even developing further and improving its new products and expanding market space for new products.

(2) Stage of penetrated innovation: Developing multi-type and serialized products

When the products of the previous stage enter the mature period, enterprises may acquire new abilities for innovation in the process of developing the product and its market. Based on this, enterprises can implement an extension of the innovation of product and technology, namely penetrated innovation, and produce multi-type and serialized products. In essence, penetrated innovation means extended innovation in accordance with the degree of technological proximity. That means that the product concepts and series extend longitudinally as well as a horizontal expansion of technology-related fields, and then product diversification can be cultivated. At this stage, when the enterprise's products face the threats that its competitors might imitate its products on the market, the longitudinal-extended and horizontal-expanded innovation can make its products and technologies get constantly upgraded, so that enterprises can make new competitive advantages and innovation capacity possible.

(3) Stage of integrated innovation: Integrating related abilities and developing new products

When enterprises have accumulated new innovation capacity in the penetrated stage, they can integrate relevant innovation abilities and develop new serialized products to achieve the transition from penetrated innovation to integrated innovation if they find new market opportunities. Integrated innovation does not mean that enterprises simply recombine various innovative elements, but they segment the new market opportunities, and then combine them with innovation capacity, so as to expand products and services, develop further new market segments and provide the clients with the value they need.

(4) Stage of breakthrough innovation: Providing the market with new value

Through the innovation practices of the previous three stages, enterprises have the chance to cultivate a matrix-type of capacity for innovation and get considerable profit, which lays the foundation for them to realize real breakthrough innovation. The so-called matrix-type capacity for innovation means the abilities for technological realization, enterprise operation, and market development regarding various types and series of product innovations. And breakthrough innovation means that the essence of the capacity for multi-type and multi-series product innovation is extracted for the formation of the most unique ability to innovate, in order to achieve new breakthroughs in technology and market as well as unprecedented market value.

(5) The progressive promotion of the innovation capacity in the four stages

The capacity for independent innovation is the fundamental factor in independent innovation. The ability which innovation subjects need in order to get unique advantages in R&D, the application of industrialization, the development of commercialization, and market promotion is the ability for independent innovation. It is an organic combination of a variety of capabilities, including the ability to control and use resources, the ability to create core technology, the ability to design and manufacture a new product body, as well as the ability to cultivate a market brand. The independent innovation capacity is not only a fundamental factor needed in order to realize independent innovation, but also an accumulating and strengthening factor in the process of moving forward towards innovation (shown in Table 4-1).

Among them, what enterprises most need in the improved innovation stage is to learn and improve their ability. However, in the process of product improvement and market development, through learning the technology of others and the conceptual development and continuous improvement of their own products, enterprises can form the field expanding ability to develop multi-type and multi-series products, so that they can realize the transition from the improved stage to the penetrated stage. In the latter stage, while the development of multi-type and multi-series products wins a chance on the market first, which may help enterprises to accumulate the ability to develop new products and make it possible for them to possess the ability to integrate the market segments and abilities to segment. Based on this, enterprises can achieve the transition from the penetrated innovation stage to the integrated innovation stage, in which enterprises integrate the market segments and abilities to segment as well as develop new products, which may make the enterprises' innovation capacity get substantial promotion, namely the formation of a matrix-type capacity for innovation as well as a featured type of innovation capacity. Once an enterprise possesses a matrix-type innovation capacity and a featured type innovation capacity, under the pressure of market competition and the interest expectation of breakthrough innovation, it will necessarily be able to make the transition from the integrated stage to the breakthrough stage.

4.2.2 The selection of the innovation mode in different stages of enterprise growth

An enterprise is like a person, which also has to experience the process from birth to death. Accordingly, if enterprises want to live longer, they all need continuous innovation, but the modes of enterprise innovation are different in different growth stages.

(1) The life cycle of enterprise growth

Figure 4-1 is the diagram of the life cycle of an enterprise. The figure is made up of three dimensions, namely innovativeness, controllability, and life cycle. In general, an enterprise's innovativeness becomes weaker and weaker and its controllability grows stronger, just like human beings. However, after the period of maturity, with the aging of the enterprise's organization and the breeding of officialism as well as the lack of internal motivation, the innovativeness of the enterprise gets weaker and the controllability problem gets worse. Unless the enterprise experiences transformation, it will not be able to find new opportunities and fields, and carry out breakthrough innovations, and get market recognition at the same time. Otherwise, the enterprise will be eliminated by the market.

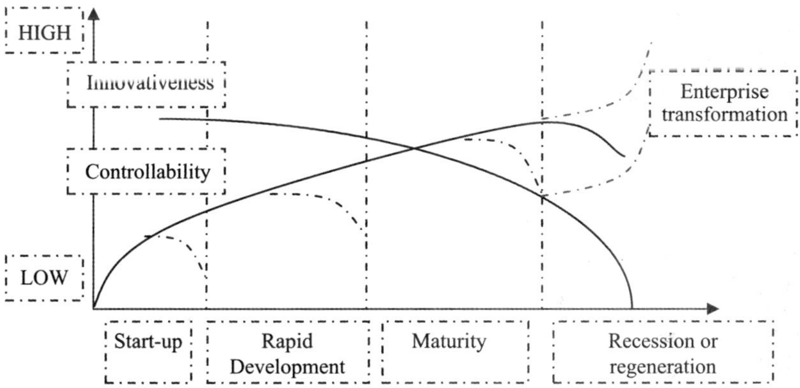

Figure 4-1 An enterprise's life cycle

In the different stages of the life cycle, enterprise innovation has different characteristics and modes. In particular, the modes of innovation in the different stages are closely related to the critical issues, the basic objective, and the core mission of enterprises (shown in Table 4-2). Enterprises can use SWOT methodology to analyze and then develop the mode of innovation in every stage. SWOT methodology means that the Threats and Opportunities which an enterprise faces in the external environment are analyzed, while their own Strengths and Weaknesses are also assessed. Based on the current situation and looking forward to the future, enterprises consequently develop their own mode of innovation.

Table 4-2 Innovation modes at different stages of the life cycle

Growth stage	Key issue	Basic goal	Main task	Mode of innovation
Start-up	How to survive	To achieve break-even and positive cash flow as soon as possible	To foster innovation capacity in some aspects	Breakthrough innovation or improved innovation
Rapid development	How to make an enterprise meet entrepreneur's expectations as soon as possible	To seize the opportunity to find a breakthrough and to achieve rapid development	To cultivate a leading business and strengthen positive cash flow and the existing capability for innovation, and strive to be the market leader	Penetrated and integrated innovation; Improved and breakthrough innovation
Maturity	How to enable the enterprise to remain on a relatively high operating platform	Sound operation and sustainable development	To seize new business opportunities, develope a cultivated capacity, and maintain the position of market leader	Improved innovation; Integrated–penetrated innovation
Recession or regeneration	How to prepare the conditions for a new development and achieving regeneration	To make the enterprise avoid recession	To find new business opportunities, and foster new capabilities, as well as striving to become the leader of a new segment of industry	Breakthrough or integrated–penetrated innovation

(2) The innovation mode in the start-up stage

In the start-up stage, an enterprise can conceive and implement at least three kinds of innovation modes through the SWOT matrix analysis chart (shown in Table 4-3). Specifically, if an enterprise is positioned in a field with weaknesses, and there are also threats from the market, it should try to avoid the weaknesses and threats and stop making efforts to carry out innovation in this field. In the field with internal weaknesses and external opportunities, it can carry out improved innovations by reducing product cost, and win market opportunities through the cost leadership strategy. In the field with internal strengths and external threats, it can only implement breakthrough innovation with the breakthrough of core technology; only in this way can it make a road for itself to win strengths on the market. In the field with internal strengths and external opportunities, it can carry out improved innovation or breakthrough innovation to win over a possible market. With such efforts, it can achieve breakeven and positive cash flow as soon as possible, while cultivating innovation capacity in some aspects. Thus, enterprises can survive and accumulate strengths in order to enter the stage of rapid development.

Table 4-3 The SWOT matrix for the selection of innovation mode in the start-up stage

	Weakness	Strength
Threat	(Avoidance)	Breakthrough innovation (core technology breakthrough)
Opportunity	Improved innovation (reducing cost)	Improved innovation + Breakthrough innovation

For example, when Huawei and ZTE were in the start-up period, they chose this mode of innovation. In the switch field with weaknesses and external opportunities, they won a market share through improved innovation by reducing costs; in the field of wireless communication products where they had advantages and there were also opportunities on the market, the two enterprises carried out breakthrough innovation from the high-end led by technology. After their products were recognized by the market, they continued to improve and perfect their products, and then got good performance on the market.

(3) The innovation mode in the stage of rapid development

After the experience of the start-up stage, enterprises usually have a certain ability to innovate and may enter the stage of rapid development with an accumulated positive cash flow. In the rapid development stage, with the SWOT matrix methodology, enterprises can also conceive and implement at least three kinds of innovation modes. That is, in the field with internal weaknesses and external strengths (shown in Table 4-4) even if an enterprise has a certain ability to innovate, it should still try to avoid the weaknesses and threats and make no effort to carry out innovation in this field. In the field with internal weaknesses and external opportunities, they can carry out improved innovation by improving product performance. In the field with strengths and threats on the market, they can only implement breakthrough innovation with the breakthrough of core technology; in the field with internal strengths and external opportunities, they can carry out penetrated innovation or breakthrough innovation; this improvement can be the improved innovation based on improving product performance as well as that based on reducing costs. With such efforts, it is possible for enterprises to achieve rapid development.

At this stage, enterprises cannot only implement improved innovation based on cost reduction and function improvement, but also breakthrough innovation with core technology breakthrough. And in contrast, enterprises should implement penetrated innovation and integrated innovation even more. Because the main task in this stage is to cultivate a position as a strong leading business and strengthen a positive cash flow and the existing innovation capacity, enterprises should strive to become the market leader. To achieve these goals, enterprises need to penetrate the field with the proximity of high technology, where they have possible weaknesses and there are opportunities outside; in the field with internal strengths and external threats, enterprises should carry out integrated innovation combining a market segment with the segmentation ability. Based on this, enterprises can grasp the opportunity to achieve rapid development and become the "excellent enterprise" expected by entrepreneurs.

Table 4-4 The SWOT Matrix for the selection of innovation mode during the stage of rapid development

	Weakness	Strength
Threat	(Avoidance)	Breakthrough innovation + Integrated innovation
Opportunity	Improved innovation (improving product function + reducing cost)	Penetrated innovation + breakthrough innovation

For example, Huawei, ZTE, and other enterprises chose this innovation mode in the rapid development stage. In the switch market, these enterprises carried out improved innovation based on cost reduction and function improvement. Particularly in the early 1990s, there were opportunities on the market, but the technology of domestic enterprises was in a disadvantaged situation and lots of core technology patents were monopolized by transnational companies, so Chinese enterprises had to pay huge patent royalties. However, Huawei had its own way to deal with this; it energetically developed "the patents at the foot of the mountain" which surrounded "the patents at the peak of the mountain" of foreign enterprises. When foreign enterprises had to use the "the patents at the foot of the mountain," Huawei exchanged "the patents at the foot of the mountain" for "the patents at the peak of the mountain." Thus, Huawei minimized its weaknesses, and grasped the opportunity to launch a large number of products from improved innovation. In addition, in the field of wireless communication equipment, Huawei and other enterprises implemented penetrated innovation and applied it to the application field. In other fields with external threats and internal strengths, Huawei and other enterprises implemented integrated innovation and gained new competitive advantages. Xunlei and Microchip are two other examples. They implemented penetrated and breakthrough innovation, and occupied the market rapidly in the field with external opportunities and internal strengths.

(4) The innovation mode in the maturity stage

In the maturity stage, the running of enterprises is relatively stable and enterprises seek further growth on a higher platform. The basic problem for an enterprise in this stage is how to remain on a relatively higher operating platform; the basic objective is sustainable development; the main task is to find and seize new business opportunities, exert the capability that is already cultivated, and maintain the position of a leader. Enterprises can conceive and implement an innovation mode in the maturity stage (shown in Table 4-5). It should be noticed that enterprises' capability for controllability and innovation are enhanced during this stage, but it is possible that the awareness of innovation could weaken.

Table 4-5 The SWOT Matrix for the selection of innovation mode in the maturity stage

	Weakness	Strength
Threat	Out of the market	Integrated–penetrated innovation
Opportunity	Improved innovation (function + cost + service improvement)	Integrated innovation + breakthrough innovation

Based on the above statement, enterprises should exit the market in the field with external threats and internal weaknesses; in the field with external opportunities and internal weaknesses, it would be best for enterprises to implement improved innovation with the improvement of function, cost, and service; in the field with external threats and internal strengths, enterprises should combine integrated innovation with penetrated innovation to implement integrated–penetrated innovation; in the field with external opportunities and internal strengths, enterprises should implement breakthrough innovation depending on their particular accumulated ability for innovation in the previous stage of rapid development; in the meanwhile they should implement integrated innovation. Here, the function of implementing integrated and penetrated innovation is to maintain the position of leadership among enterprises; the function of breakthrough innovation is to prepare enterprises for implementing the next phase, transformation and regeneration. Typically, China has a lot of manufacturing LMEs which are in the maturity stage. The main mode of their independent innovation is precisely the combination of integrated and penetrated innovation, while making efforts to implement breakthrough innovation.

(5) The innovation mode in the stage of recession or regeneration

After the maturity stage, the enterprise may enter the next stage, in which there is either recession or regeneration. If enterprises implement breakthrough innovation in the maturity period, most enterprises would get regeneration. Once an enterprise achieves regeneration, this enterprise may enter a new period of rapid development. Thus, in this situation, enterprises should decidedly exit the field, where there are external threats and internal weaknesses; in the field with external opportunities and internal weaknesses, enterprises should fight to win or die and industriously push through new breakthrough innovation; in the field with external threats and internal strengths, implementing integrated–penetrating innovation may be beneficial; in the field with external opportunities and strengths, enterprises should implement integrated–breakthrough innovation. Based on this portfolio mode of innovation, enterprises could really move towards regeneration (shown in Table 4-6).

Table 4-6 The SWOT Matrix for the selection of the mode of innovation in the situation of enterprise regeneration

	Weakness	Strength
Threat	Out of the market	Integrated–penetrated innovation
Opportunity	Radical innovation	Integrated–breakthrough innovation

If enterprises do not implement breakthrough innovation in the maturity period, most of them may fall into recession. Once it happens, it means that they can only prolong their last gasp. Thus, enterprises should decidedly exit the field with external threats and internal weaknesses; in the field with external opportunities and weaknesses, the implementation of penetrated innovation is the only choice; in the field with external threats and internal strengths, the implementation of integrated innovation is the only choice; in the field with external opportunities and internal strengths, enterprises should implement integrated–improved innovation. Based on this portfolio mode of innovation, enterprises may naturally prolong their survival period (shown in Table 4-7). Typically, a lot of Chinese enterprises which are in the situation between recession and regeneration implement this kind of portfolio mode of innovation.

Table 4-7 The SWOT Matrix for the selection of innovation mode in the situation of recession

	Weakness	Strength
Threat	Exit the market	Integrated innovation
Opportunity	Penetrated innovation	Integrated–improved innovation

4.3 The construction of the trajectory of and ability for industry innovation

4.3.1 The selection of a generalized trajectory of industry innovation

(1) The selection of the trajectory of innovation for different types of industries

When different countries carry out industry innovation, they will naturally select corresponding trajectories of innovation according to the industry types. For example, in the emerging industries, advanced countries with advanced technology will start from the original innovation and promote overall industrial development through breakthroughs in core technology. However, to develop new industries, most developing countries will introduce mature and advanced technologies from developed countries, followed by implementation, adoption, and re-innovation. For the industries with strong technology correlations, both developed and developing countries will first implement the mode of integrated innovation to improve the capacity for independent innovation. For the industries which play a key role in national economic development, starting from original innovation is the common choice for most countries. However, due to their relatively backward technology, it is necessary for developing countries to introduce and adopt advanced technologies, and grasp the existing technologies. Of course, if developing countries only focus blindly on the introduction of technology, and do not focus on the implementation, adoption, and re-introduction of imported technologies, it will form a kind of dependence on technology of developed countries to some extent. Naturally, it is not conducive to the stable development of related industries in developing countries.

(2) The selection of a trajectory for industry independent innovation in the different stages of development

Industrial development will go through the process of "inoculation—formation—rapid development—maturity—recession." There are different trajectories for industry innovation in the different stages of industrial development. For example, in the inoculation period, industrial development experiences great uncertainty and has a weak foundation. Adopting the "technology introduction—implementation and adoption" trajectory for innovation is beneficial to the cultivation and development of an industry. In the formation stage, the gradual development of an industry will cause the industry scale to increase gradually, and technological development paths are various; the "technology introduction—implementation and adoption—imitation and innovation" trajectory of innovation is conducive to the rapid growth of industry. In the period of rapid development, industrial development has good prospects, with the increase of industry profits and investment attractiveness, and adopting the "technology introduction—implementation and adoption—imitation and innovation" trajectory of innovation and "technical accumulation and technological transformation" or the "promoting original innovation" strategy can help to foster an industry's capacity for independent innovation. In the maturity period, the industry pattern has basically shaped; the profit rate of industrial investment becomes stable. It is difficult to carry out innovation based on breakthrough of the existing technology. The ability for independent innovation should be cultivated through "original innovation"; this helps a country maintain certain advantages in international competition. In the recession period, the profit rate of industrial investment decreases, the survival potential of an industry can be exposed only through "integration innovation," and a new industrial development path can be created through "original innovation."

4.3.2 The selection of a trajectory for industrial independent innovation: Taking the Chinese equipment manufacturing industry as an cxample

The selection of a trajectory for industry innovation is greatly related to the type of industry. Thus, the Chinese equipment manufacturing industry will be analyzed as an example.

(1) The equipment manufacturing industry and its characteristics

Technical equipment is a knowledge-, technology-, and capital-intensive product. The equipment manufacturing industry has characteristics such as the high-tech content, a complex production process, high-tech requirements, completeness, a complex relationship for coordination, high correlation, etc. It is representative of the national capability for integrated manufacturing; and the ability for the research and manufacture of major technical equipment is an important indicator to measure the level of industrialization and comprehensive national strength. It is also the important means and industrial carrier of the transformation from high S&T to practical productivity. In the process of a country's industrialization, the equipment manufacturing industry plays a role in equipping the national economy. Within the industry, the major technical equipment manufacturing is the core, and it is the mothership of the entire equipment manufacturing industry with the highest level and technical content, the most complex and widest coordination as well as the most difficult system of integration.

(2) The trajectory for independent innovation of the localization of the Chinese equipment manufacturing industry

There are three conditions in the innovation and development of the equipment manufacturing industry.

Condition one: Introduction—falling behind—re-introduction—falling behind again. After enterprises have introduced the first-generation of some kind of foreign equipment, if enterprises produce the first equipment after implementation, adoption, and improvement without government support or related market opportunity, the first equipment or the first set of equipment will not be used in downstream industries (demand side) due to its popularity, reputation, and system (except for quality). Then the first equipment will die on the vine, and this will inevitably lead to the introduction of the second-generation of advanced foreign equipment. If the situation cannot be improved, China will inevitably fall into the cycle of "introduction—falling behind—re-introduction—falling behind again." Consequently, this kind of equipment will never achieve localization in China. At present, lots of equipment manufacturing industries in China are in this embarrassing situation, such as the direct air-cooled equipment for power stations.

Condition two: Progressive catching-up. The difference between this condition and the previous one is that we focus on the practical application of the first independently developed equipment after the implementation, adoption, and re-innovation. After further performance optimization, the equipment may enter the stage of localization and mass production. If we focus on further innovation based on mass production along with the accumulation of experience and further adoption of foreign technologies, we can make the local equipment reach an advanced international level, such as the introduction and re-innovation of Chinese thermal power equipment.

Condition three: Leapfrogging innovation. With the accumulation of the innovation capacity of Chinese equipment manufacturing enterprises, we can directly introduce or buy advanced foreign technologies through acquisitions or joint design. Combined with the actual situation, further design and optimization as well as the strengthening of trial-manufacture could make it possible to achieve localization quickly, such as the research and manufacture of Chinese aerospace equipment.

4.3.3 The inherent structure and construction of the industrial capacity for independent innovation

(1) The inherent structure of industrial capacity for independent innovation

Industrial capacity for independent innovation is the comprehensive capability where industrial subjects take advantage of means of innovation, such as technological innovation, management innovation, market innovation, etc., in order to achieve a breakthrough in key industrial technology, form independent intellectual properties and industrial standards, create the international competitiveness of industry and promote advantages in industrial competition in the process of original innovation, integrated innovation, and the innovation of introduction, implementation and adoption, and re-innovation. Industry's capacity for independent innovation is not the simple aggregation of enterprises' abilities for innovation; it is also related to the collaborative mechanism of innovation networks which are made up of intra-industry enterprises and other innovation subjects.

(2) The proactive construction of industry's capacity for independent innovation

Based on the foregoing discussion and the experience of Japan and the Republic of Korea, the proactive construction of an industry's capacity for independent innovation is shown in Figure 4-2. With the support from a particular industrial development environment and the collaboration of industry innovation networks, through enough investment in industrial innovation resources, such as labor, capital, infrastructure, and basic technology, taking the combined mode of introduction, adoption and innovation, and re-innovation, integration innovation, original innovation, we may form independent intellectual properties, and develop technologies and products as well as standards in the corresponding industries; and this industry's ability for independent innovation can also be cultivated proactively in the framework.

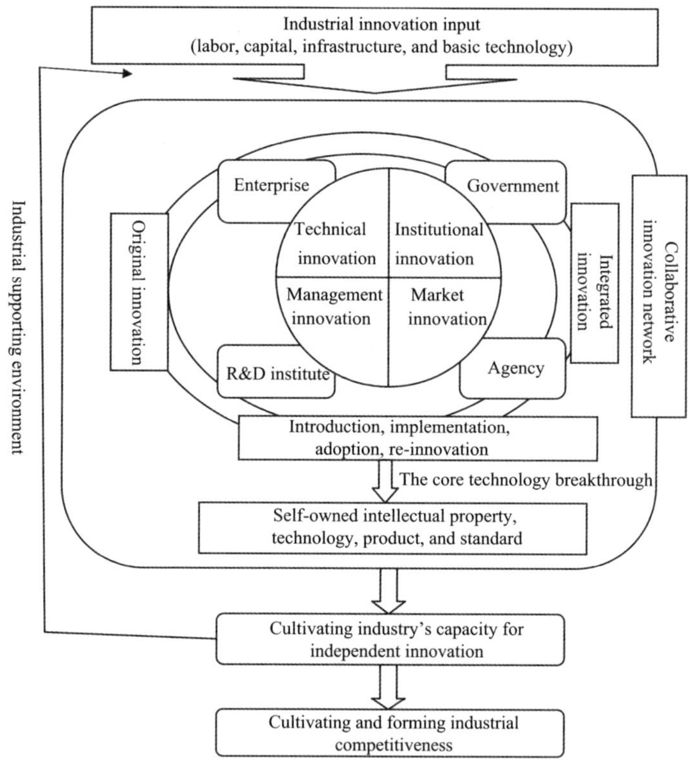

Figure 4-2 The proactive construction process of industry's capacity for independent innovation

References

Burgelman, R.A., Maidigue, M.A. & Wheelwright, S.C. *Strategic Management of Technology and Innovation*. New York: John Wiley, 1995.

Dosi, G. Technological paradigms and technological trajectories: A suggested interpretation of the determinants and directions of technical change. *Research Policy*, 1982, 11 (3): 147-162.
Furman, J.L., Porter, M.E. & Stern, S. The determinants of national innovative capacity. *Research Policy*, 2002, 31 (6): 899-933.
Guo, J. & Zhu, K. On concept, meaning and component parts of independent innovation. *Productivity Research*, 2008 (2): 16-17, 38.
Lall, S. Technological capabilities and industrialization. *World Development*, 1992, 20 (2): 165-186.
Lu, F. Understanding of independent innovation. *Science and Technology Industry of China*, 2006 (10): 42-45.
Mao, Y. & Wang, J. Research on the path of independent innovation based on product upgrading. *Management World*, 2006 (5): 114-120.
Porter, M. *Competitive Advantages*. Beijing: Huaxia Publishing House, 2002.
Shi, Q. & Shang, Y. *China Industry Development Report 2001*. Beijing: China Light Industry Press, 2002.
Song, H., Mu, R. & Ren, Z. Study on indigenous innovation and the measurement of indigenous degree of innovation. *China Soft Science*, 2006 (6): 60-66.
Wang, L. et al. Research on innovation model selection and mechanism of independent innovation ability promotion. *Science & Technology Progress and Policy*, 2006 (7): 45-51.
Wang, Y. China's economic development faces a major adjustment. *China Soft Science*, 2005 (7): 7-9.
Westphal, L.E., Rhee, Y.W. & Pursell, G. Korean industrial competence: Where it came from. *World Bank Staff Working Papers*, 1981: 469.
Xia, Z. & Yuan, J. Systematic building and components of enterprise independent innovation ability. *Industrial Technology & Economy*, 2007, 26 (10): 2-6.

Chapter 5

The method and mechanism of technology integration in enterprise independent innovation

5.1 The relative concepts of technology and technology integration

5.1.1 Definition of technology

According to the definition in *Merriam-Webster's Collegiate Dictionary* (1995), technology can be divided into four categories: i) The practical application of knowledge especially in a particular area; ii) a capability given by the practical application of knowledge; iii) a manner of accomplishing a task especially using technical processes, methods, or knowledge, as well as the scientific method for a practical purpose; iv) a specialized aspect of a particular field of endeavor. In specific research, scholars divide technology into narrow technology and general technology.

(1) Definition of narrow technology

Seeing technology from the perspective of an enterprise, Holatius (1995) claims that technology has three levels: Product, process, and people; Katz et al. (1996) divide technology into four types, including general information, special information, procedure and operation, as well as hardware.

Robinson (1990) defines technology from two dimensions. He thinks the first dimension is the horizontal composition of technology: i) The inalienable core technology or peripheral technology; ii) permission to use the right or knowledge, especially the technology license; iii) hardware, including the typical equipment of technology or other intermediate products and the final products; iv) software, such as files or verbal communication which is non-materialized technology. The second dimension is the vertical classification of technology: i) consumer technology (the skills related to the use of technology); ii) product adoption technology (the skills for using technology under specific conditions); iii) production technology (skills for copying technology); iv) design improvement (the skills needed to improve the design of product or process); v) design technology (the skills needed to design a new product or process).

Seeing technology from the perspective of manufacturing, Xia Dawei et al. (2002: 3) divide technology into three categories: i) Product technology: as the primary link of the formation of particular products, it not only includes the integration of the technologies of research, development, design, testing, etc., which are related to the forming of the particular products, but also involves the physical technologies reflected by the principles, structures, and properties of the particular products; ii) equipment technology: it is mainly reflected by the specification, performance, and quality of equipment production; iii) process technology: it is mainly the integration of the links of the design of the production process and the relevant homemade processing equipment, and it is concretely reflected as a series of rules or knowledge which are related to the production process, to the operation methods, to the quality control, etc.

(2) Definition of general technology

Some scholars understand technology from a broader perspective; they divide technology into three levels: i) Technological process, processing method, labor skills, and know-how which are developed according to the principles of natural science and the experience of production practice; ii) corresponding production tools and other supplies and equipment which make these methods, processes, skills, and know-how come true; iii) knowledge experience and methods which are applied to the application of the modern division of labor and the scale of production and other requirements for an effective organization and management of all resources in the production system (Fu Jiaji, 1998: 12). Obviously, the general understanding of technology includes not only narrow technology, but also organizational and managerial skills.

It is worth noting that the technology referred to in "technology integration" accepts the general concept, which mainly includes four aspects: i) Product technology: including material design, structural design, appearance design, software, etc., as well as the matching ability and carrier of these technologies; ii) manufacturing technology: including process and equipment technology as well as the corresponding ability and carrier of these technologies. Process technology includes the development of process, design of parameter, and control technology; equipment technology includes tooling and equipment development and preparation; iii) individual skills, including experience, professional knowledge and skills, operational skills, personal relationships, and the cognition of enterprises and the external environment; iv) organization skills, including the ability and conventions of enterprises in supplier management, outsourcing management, knowledge management, project management, external relationship management, and market management as well as other aspects. The four kinds of technologies above can be divided into tangible and intangible technologies from the perspective of the form, and internal and external technologies according to the source.

5.1.2 Two kinds of technology integration

(1) The technology integration facing product

Technology integration was first proposed by Iansiti of Harvard Business School (HBS) in 1993. Iansiti claims that technology is the method of evaluation, selection, and extraction of a new product and its process. Its purpose is to match enterprises' all kinds of knowledge with the special environment of technology selection and decision-making to successfully launch a new product or process (Iansiti, 1993, 1995, 1997, 1998).

Iansiti's research on technology integration mainly focuses on the industries of semiconductor, personal computer, and software in America and Japan. The technical characteristics of these industries include complicated technology, rapid and discontinuous development. The concept of a product or a technology can be achieved by many technological proposals. There are very subtle relationships among different technologies and between the enterprises' actual situation and different technologies. The selection of technology will result in a difference in the development of the performance of the product. The time needed for product development is an important competitive factor. Many companies lose their leading advantages due to the point of technology discontinuity; some even become losers. Therefore, in this industry, for the development of product, it is very important to choose a technology suitable for an enterprise's practical environment.

Iansiti's research on technology integration stresses the selection of product technology. It stresses the match of the proposal of the selected product and process with the enterprise's actual environment (which includes the system of the manufacturing and production, the ability of the supplier, etc.) in order to achieve the manufacturability of the product (shown in Figure 5-1). It is a new R&D management concept and method (Iansiti, 1993) called the "technology integration facing product."

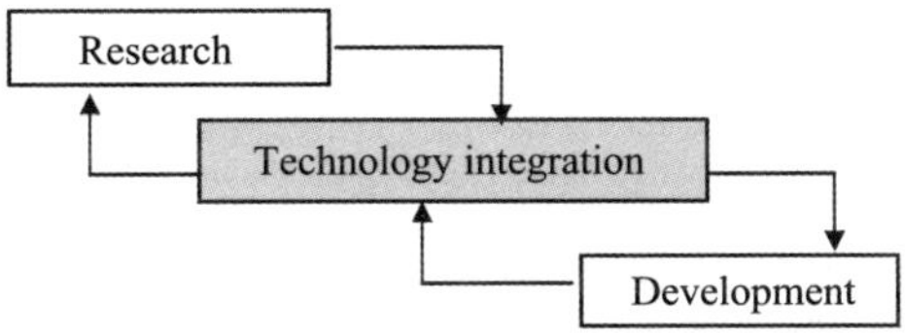

Figure 5-1 Technology integration (facing product) by Iansiti

(2) The technology integration facing industrial production

From the perspective of enterprises, technology includes product development technology and manufacturing technology. The former proposes the concept and the ability to create a new product. The latter is the ability to produce products attractive in price and quality on a large scale based on the above concept. The previous domestic research on technological innovation paid more attention to the achievement of product technology, but less attention to how to achieve industrial quantitative production. The technology integration facing industrial production stresses the transfer from technology in the laboratory to industrial production, which means that enterprises get the proposal for the product and the production technology, which matches the enterprise's business environment in order to carry out a series of activities in industrial production through technology selection, sourcing, and extraction. The business environment refers to the market, industrial policy, industrial production environment, etc.

Certainly, technology fusion is different from technology integration. Technology fusion analyzes innovation from the perspective of technology; it is the combined application of product and process of different categories of technology (narrow technology); meanwhile, technology integration studies technological innovation from the perspective of management; it is the organic integration of technology resources with different properties (it refers to general technology, including both tangible tools and equipment and intangible operation methods as well as organizational managerial skills, etc.) and commercial resources (shown in Figure 5-2).

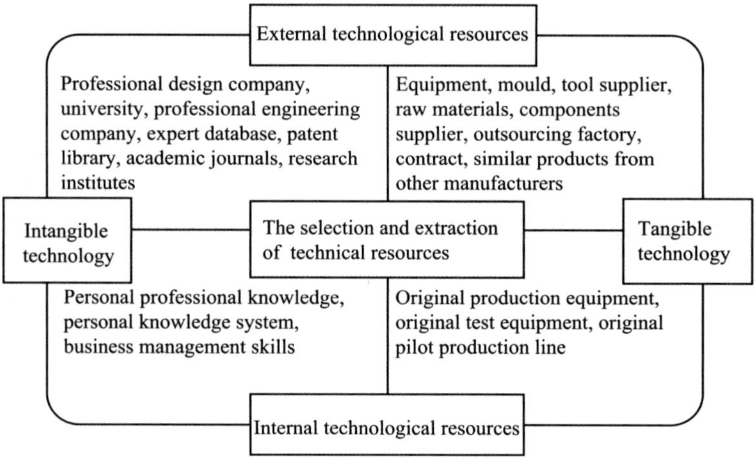

Figure 5-2 The essence of enterprise technology integration

According to the survey regarding Chinese enterprises, the ability and environment of most Chinese enterprises are different from those in Iansiti's research (Deng Yan & Lei Jiasu, 2006). Meanwhile, the technology integration of Chinese enterprises is very different from that of overseas enterprises.

First, the novelty of technology. The product Iansiti has researched is highly novel in technology. Many enterprises start from basic research or applied research, having great difficulties and uncertainties in technology development. However, many Chinese enterprises do not engage in basic research. Their innovation starts directly from applied research or production design. Experiences of other enterprises can be regarded as references (for instance, after an overseas enterprise releases a product, domestic enterprises will develop a similar product for the home market). Hence, there is little difficulty in the selection of basic technology and prototype technology trajectory in the earlier stage.

Second, source of prototype technology. In Iansiti's technology integration research, the prototype technology of a product mainly comes from the enterprise's internal research institutes. However, in the innovation process of Chinese enterprises, except for internal R&D, the employment of external technology is a very important resource of prototype technology (which includes introduction and imitation). How to select and integrate external resources effectively is a common problem for enterprises.

Third, cost pressure. Many core technologies and parts of domestic products are controlled by overseas enterprises. Chinese enterprises need to pay a high price for technology transfer or core parts purchase. Room for profit is seriously decreased. What's more, in order to survive on the market which is full of famous international brands, the products of many Chinese enterprises come into the medium- and low-end market. Besides, because of the violent homogeneous competition on the domestic market, product cost is the key point for the success of innovation. In the process of product design and the development of manufacturing technology, enterprises have to balance the cost, performance, and efficiency.

Fourth, the restriction of industrial supporting ability. Chinese basic industries are weak, such as in special materials, equipment manufacturing, precise instruments, precise components, and parts and mould processing. Lots of excellent technologies cannot be applied in industrial production because of the lack of suitable suppliers of materials and old-fashioned equipment manufacturing technology. Most of them stay in the stage of laboratory achievement.

5.1.3 The relationship between technology integration and independent innovation

(1) In the following two situations, enterprises need to implement the technology integration facing industrial production

The technology integration facing industrial production takes achieving industrial mass production of new technology products as the goal. New technology is relative to the existing technology of enterprises, including nascent technology and the technology which other enterprises have used but the enterprise itself still does not use. In the following two situations, enterprises need to implement technology integration facing industrial production, namely seeking the appropriate convergence point among the new technology, the ability of the enterprise, the market, the supporting resources, and the policy environment.

The first situation: After the development of the new technology or the product concept, if there is no ready-made industrial manufacturing plan or business mode which can be used as a reference, enterprises need to further develop product technology and manufacturing technology. Mass production is a great challenge for the commercialization of nascent technology. Because there were no ready-made equipment suppliers, the main obstacle in the process of the commercialization of Xerox and Sharp diode technology was the inability to develop and manufacture production equipment. Besides the departments of product and process engineering, Sony set up the special department of manufacturing engineering which was in charge of the development of tools and equipment. In the end, Sony was successful in the commercialization of diode technology (Wood & Brawn, 1998).

The second situation: When adopting external technology, enterprises do not completely copy the product scheme or manufacturing solutions of other enterprises. They need to adjust or further develop the external technology according to their own situations. For example, when BYD entered the nickel-cadmium (Ni-Cd) battery field, they did not introduce the complete set of Japanese automatic production line. According to the characteristics of the domestic manufacturing environment and each part of the production line, BYD solved the key technical problems through technology integration; to reduce the cost of manufacturing, BYD took advantage of the manpower maximally in many aspects, excluding the necessary machinery and equipment. Moreover, they developed key equipment independently, thus finally achieved cost advantage.

(2) Technology integration is an important link in technological innovation

Technological innovation is a business concept, which comes from the understanding of the commercial potential of technology. The concept means the final transformation of the potential into a commercial product (Fu Jiaji, 1998). The prototype technology and the sample are not the ultimate goals of technological innovation; only when the enterprises turn the prototype technology into an industrial product and push the product onto the market through technology integration, does it mean completing technological innovation. Therefore, technology integration is an important link in technological innovation. However, seen from the enterprises' point of view, as long as enterprises use new technologies, the process belongs to the category of technological innovation, according to the *Oslo Manual* (2005). Accordingly, technology integration can be internal or external R&D activities (shown in Figure 5-3), and it can also be the partial introduction based on mature external technology (shown in Figure 5-4). Based on all of the above aspects, it is not difficult to see that technology integration facing industrial production is the substantial process of new technology which transforms technology from an achievement to industrial production. And then, it is not difficult to judge that technology integration facing industrial production is one of the core issues of enterprise technological innovation.

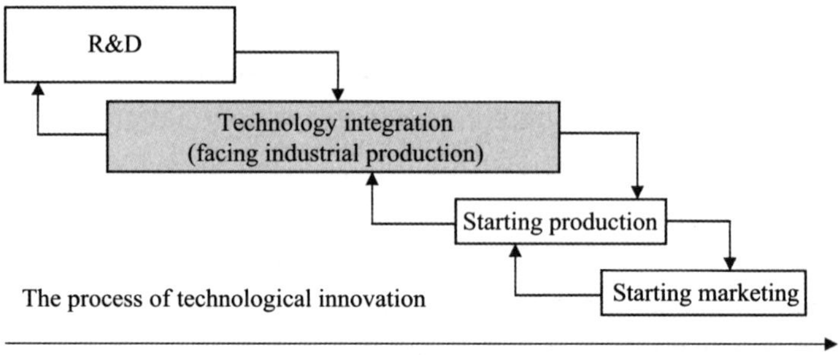

Figure 5-3 Enterprise technology integration and technological innovation based on R&D

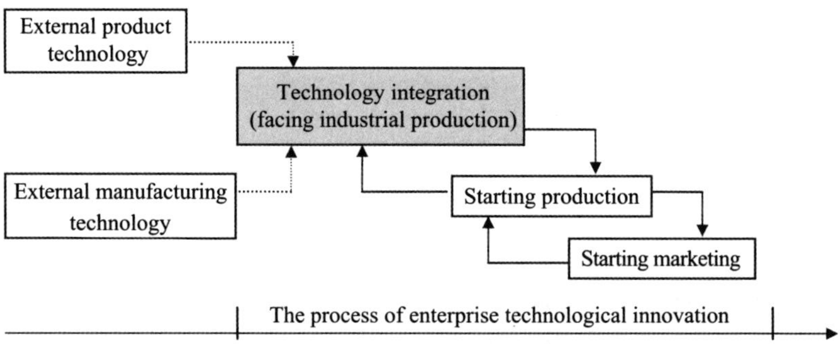

Figure 5-4 Enterprise technology integration and technological innovation based on the introduction of technology

In some cases, technology integration facing industrial production is also the necessary link in the process of enterprise technological innovation. The technological innovation of enterprises can start with the exploration of new areas, the use of the existing technology or mature external technology. In these three kinds of innovation modes, enterprises need to combine new technology with their own ability and industrial environment through the technology integration facing industrial production to realize the commercial value of technological achievements.

After investigating the innovation practice of some enterprises, we find that enterprises need to fill in three gaps in the process of the transformation of technological achievements from the R&D platform to the production platform: i) The transformation of prototype technology into available, manufacturable, testable, and documentable product technology; ii) the development of manufacturing technology, which can copy samples with low cost and high efficiency; iii) the realization of the industrialization of the manufacturing plan, including equipment development and manufacturing, tooling preparation, personnel preparation, etc. Enterprises may encounter a variety of problems and difficulties in every step. In many cases, when encountering some difficult problems, enterprises may ultimately fail to fill in one of the three gaps, which results in the failure of commercialization of the new technology.

In addition, with the acceleration of technology development and the promotion of its complexity, it has become a trend for enterprises to make full use of external technical resources in order to enhance the internal capability of technological innovation. The essence of technology integration facing production is that enterprises integrate a variety of internal and external technical resources that the enterprises need around technological innovation. Now, many studies emphasize the construction of enterprise R&D capability and investment. However, for many domestic SMEs, with a lack of capability for technology R&D, if enterprises can organically combine the various external technical resources, especially the defective technical resources, with the existing internal resources, they can also achieve technological innovation. For the enterprises which have the ability to carry out R&D, when applying internal R&D achievement to industrialized production, they must also have a certain ability for technology integration facing industrial production. Therefore, enterprises should strive to become an effective technology integrator, which plays an important role in the process of enterprise innovation.

(3) Technology integration is an important mechanism and method for helping to succeed in independent innovation

From the perspective of the innovation value chain, a country's capacity for independent innovation should be a synthesis of six aspects:

i) The ability to explore new knowledge; ii) the ability to acquire and develop technology; iii) the ability to integrate technology; iv) the ability to manufacture; v) the ability to support industry; vi) the ability to develop new markets. Without any one of them, its independent innovation will be limited. At the enterprise level, the capability for technology integration contains the ability to take advantage of various internal and external innovation resources, including the ability to mobilize and allocate internal resources, the ability to search for and acquire external resources, the organizational and managerial abilities concerning the procedure, and the technical ability of a program.

The capability for technology integration of enterprises facing industrial production is the guarantee for enterprises, university, and research institutes to transform technological achievements into practical productivity. Especially, for most innovation resources are distributed outside the enterprises in China, enterprises are required to integrate more superior resources controlled by different subjects, for the formation of the joint forces of enterprises' independent innovation. At the same time, foreign advanced technology is also an important resource which domestic enterprises need to integrate into the process of innovation. In the process of integrating foreign technology resources, whether Chinese enterprises can effectively acquire and integrate these resources is a key for them to realize independent innovation. Based on this, in the process of the independent innovation of enterprises, in addition to the building up of the R&D capability, more attention should be paid to enhancing enterprises' ability to carry out technology integration.

5.2 The content and process of technology integration

5.2.1 The content of technology integration

Regarding content, technology integration has three aspects: Technology selection, technology sourcing, and technology extraction.

(1) Technology selection

Technology selection refers to the comparison of a variety of available technical resources; its purpose is to make the selected technologies (including external and internal technology) match the business environment (including policies, market, and industrial environment). The gist of technology selection includes technical advancement, applicability, etc. The standards and methods for comparing various technologies directly affect the effect of technology integration.

1) Technology advancement and applicability. The most advanced technology is not always the best choice. Ampex in the United States first launched the commercial video recorder onto the market in 1956. It selected the four-head scanner technology with excellent technical performance, but also with high costs and complex manufacturing. Ampex's VCR had a high price, mainly on the commercial market. The Japanese company, Sony, got part of the video technology through a technology exchange with Ampex and initially adopted the four-head technology. Sony later became aware of the huge market potential of the household VCR, but the price had to be reduced in order to open up the market. Hence Sony gave up the four-head technology, and turned to double-head technology, which is not as good as the four-head in technical performance, but is easier to manufacture and costs less. In 1962, Sony launched the double-headed video product. Through continuous efforts, in the 1970s Sony launched the double-headed VCR, Betamax, at a price that consumers could afford, thus starting to become the dominant enterprise in the VCR industry.

2) Technology selection and business environment. How to guarantee a good match to the business environment is a key to the selection of technology. T Company is a domestic biological pharmaceutical company. Most of its technologies originate in the laboratories of foreign research institutes. Its internal R&D department does not engage in research work; instead its main function is to evaluate and select the technology provided by other research institutes, manage and coordinate the whole process of turning the new technology from the external research institutes into the company's pilot plants and manufacturing workshops. The technical personnel in the R&D department have rich experience in production and marketing. Apart from selecting technology from the perspective of technology assessment, they focus especially on evaluating technology in terms of commercialization and feasibility, such as market prospects, production feasibility, etc. In addition, T Company would like to introduce technology from the research institutes with which it has a long-term relationship of cooperation. Being involved in the laboratory research stage helps T Company guarantee a good combination of technology selection and business environment.

3) Technology selection has been divided into two kinds. One is screening in a large number of schemes through experimentation and testing; the other is prescreening by experience, and then testing a few schemes prescreened. For example, American enterprises usually select technology depending solely on the tests, but Japanese enterprises rely more on the experience of previous projects. This also leads to different selections of technology by the enterprises of the two countries to some extent: Advanced technology and improved technology (Iansiti, 1997). Usually, the testing methods include computer-aided simulation, test room, the trial production workshops of the R&D department and the production department as well as that of commercialization, which are in an increasing sequence in terms of the representativeness of the industrial production environment (Pisano, 1996). Brown and Duguid (2000) believes that when we select technology, we should choose appropriate technology. It is not always true that the production technology with a higher degree of automation has higher productivity and more stable quality.

(2) Technology sourcing

Technology integration is not only the integration between different types of technologies, but also the integration of the goals, interests, and resources of every participant. Technology sourcing mainly refers to the acquisition of external technology and resources which enterprises need and the process of communication and interest coordination among the participants. Technology sourcing and technology acquisition are different. Technology acquisition is a broader concept, including technology sourcing and internal development. Jonash (2001) thinks the technology sourcing modes can be divided into two kinds: One is explicit technology sourcing, namely technology sourcing through formal technology license agreement, R&D commission contract, and technology suppliers, as well as the personal relationships with universities, governmental laboratories and technical personnel; the other is implicit technology sourcing, which means enterprises realize technology sourcing by active or passive cooperation with parts suppliers through joint ventures, strategic alliances, and global subcontractors system. Some scholars also divide technology sourcing methods into technology purchase, imitation, alliance (quasi internalization), mergers and acquisitions (M&A) (internalization) with the purpose of technology sourcing, and so on. Among them, technology purchase means the purchase of a patent, the right to use a trademark, R&D achievements, personnel training, design drawings, and equipment that enterprises need. Technology alliance includes the following forms: OEM, technology license, technology exchange, and R&D alliance, as well as joint venture. At the same time, the Korean scholar Kim Linsu (1998) summarized four kinds of technology sourcing (shown in Table 5-1).

Table 5-1 Technology sourcing modes of Korean enterprises (Kim Linsu, 1998)

	Active offer of foreign suppliers	Passive offer of foreign suppliers
Market medium	Formal mechanism: Direct investment, foreign license, turnkey project, consulting	Trade by commodities: Standard machine
Non-market medium	Informal mechanism: Foreign buyers, distributors' technical help	Informal mechanism: Decomposition research, inspecting trade, academic journal

Concerning the advantages and disadvantages of the different methods of technology sourcing, Chen Song & Feng Guoan (2003) hold the opinion that technology purchase is applicable to the sourcing of embodied technology with few implications and coded "documentation" technology. Due to the short time of face-to-face communication for both sides of technology trades, enterprises are often unable to import implicit "process" technology, such as know-how, experience, and skills. However, technology alliance members have a better chance of communicating and cooperating face-to-face for a long time. They can not only import explicit documentation technology, but also implicit knowledge (know-how, skills, and experience), which can cultivate the ability of organizational learning and improve the enterprises' accumulation of knowledge and technology; through technology M&A, enterprises can quickly realize the whole sourcing of core technology and have a high degree of control over target technology. Accordingly, in recent years, merging and acquiring high-tech enterprises with patent technology has become an important choice for some developed and emerging countries and regions in order to rapidly acquire the technology and innovation resources which are closely related to the core technology of the merged and acquired companies.

(3) Technology extraction

All kinds of selected knowledge and technology must eventually materialize into a new product and production system. Technology extraction is the process of the extraction and combination of selected technologies, and reflect in the forms of technical documents, samples, and production lines. From the perspective of the transformation of knowledge, technology extraction is actually the process whereby enterprises combine, extract, and encode explicit and implicit knowledge, which is scattered in various technical resources (including all the departments/enterprises and individuals) into knowledge of product and manufacturing. Among them, one is the extraction and coding of explicit knowledge, and the other one is the organic combination of knowledge and technology scattered throughout the different departments and organizations.

5.2.2 The process of technology integration

Considering the process, technology integration focuses on the phases from laboratory achievement to industrial mass production. Analyzing from the perspective of technology development, technology integration has three stages.

(1) Three stages of technology integration

Productization—the selection, sourcing, and extraction of product technology. The main function of the product prototype is to verify the feasibility of the technological hypothesis with little consideration of the appearance, service condition, raw materials, and technological structure of the product. The main task in the productization stage is to turn prototype technology into available, manufacturable, testable, and documentable technology, namely the industrialization design of the product. In addition to the detailed drawings of design and process, enterprises should be clear about product performance, test standards as well as the type and source of the raw materials, and verify and amend the design project by testing samples. Generally speaking, after several rounds of "design—pilot production—modification—pilot production," the product's project design can be finished. The productization stage will not be finished until having passed a pilot production.

Engineering—the selection, sourcing, and extraction of manufacturing technology. The purpose of the productization stage is to verify the feasibility and rationality of product technology by means of a small amount of samples, including the function, performance, reliability, manufacturability, cost, etc. However, industrial production means producing products with stable quality at low manufacturing cost and high efficiency. The main task in the engineering stage is to develop the manufacturing technology for mass production, and establish production and quality control procedures, methods, and standards. Specific tasks include the design of the technological process and parameter, the development of operation instructions, the scheme design of quality control, and the design of equipment, tooling, and mold. Similarly, enterprises can test and modify manufacturing technology through a trial production. Generally speaking, after several rounds of "engineering design—trial production—modification—trial production," the product's project design can be finished. The engineering stage will not be finished after having passed a trial production.

From product technology to industrial production, there are four solutions as follows: i) Introducing a whole set of existing manufacturing solutions from other enterprises; ii) exploiting external manufacturing capability (OEM); iii) the internalization of external manufacturing capability (M&A); iv) the independent development of manufacturing technology or the adjustment of the original production system. In addition to the first method, the last three belong to the category of technology integration.

The establishment and adjustment of the manufacturing system. The ultimate task of technology integration is to integrate products and manufacture technology in the actual business environment and set up a production system with efficient operation. The final stage is the establishment of a production system and mass production, including tooling, mold, and equipment installation and debugging, staff training, etc. Before finalizing the product design, enterprises need to adjust the products and the manufacturing projects according to the actual situation. For example, enterprises need to consider changing the suppliers of the raw materials in the design scheme from the perspective of lower costs. By optimizing the process parameter, enterprises can improve the production efficiency of mass production in the future.

(2) Technology integration of different industries

Dividing technology integration into different stages does not mean it is a linear process. Instead, each stage in the process of the actual operation is overlapping, sometimes interlaced with other stages. The stage division is for convenience and further study. In each stage, there is a process of technology selection, sourcing, and technology extraction. According to the characteristics of its production process, the product can be divided into a process-oriented product and a processing-and-assembly-oriented product (including process assembly type and discrete assembly type). Although the two kinds of new technologies both require the process of technology integration from the test room to industrial production, the specific operation process is different.

The production process of the assembly product is mainly the assembly and testing of the various components and parts; most of the manufacturing work has been done by the suppliers of the upstream components and parts. Finding the appropriate supplier of the components and parts becomes the key task of technology integration and the task needs to be finished in the productization stage. In addition, the design of the process structure of the product determines its manufacturing process to a great extent. Therefore, the productization stage is significantly important for the entire process of technology integration.

The performance of the process-oriented product is closely related to its manufacturing technology. From the sample in the test room to the industrial product, this product does not have the sole productization stage. Instead, the source and proportion of the raw materials in the laboratory will be adjusted in the scale-up process according to the actual situation. The industrial production process is not only a simple amplification of the laboratory manufacturing process; sometimes it is hard to be achieved. Therefore, the engineering stage plays a decisive role in the success of technology integration.

5.2.3 The problems enterprises face in the process of technology integration

(1) The transformation of technology from the laboratory to the market

From the studies on the transformation from laboratory technology to commodity, we can see the necessity of technology integration. Fleck (1994) thinks that innovation is the allocation of sources of technology, organization, and capital to make an effective system. In the process of innovation from prototype technology to the market, there are many uncertainties, e.g., whether the product design can be economically sustainable, whether the product could be accepted by the market, and so on (Parthasarthy & Hammond, 2002). Brown & Duguid (1996) divides the new product development process into six stages: Generation of the idea, preliminary screening and analysis, preliminary design and development (carrying out the idea), testing of the prototype, final design (facing mass manufacturing), and mass manufacturing. Among them, the design stage facing mass manufacturing is the so-called technology integration stage. Brown & Duguid (1996) think that the reason the technological innovation carried out by numerous companies fails in the design stage facing mass manufacturing is that they cannot transform the one-off design scheme (prototype) into the mass production scheme. For example, Xerox's Palo Alto laboratory in the United States developed a computer with a mouse in the 1970s, but the company did not push the mouse onto the commercial production stage (Brown & Duguid, 1996, 2000). The study of Nevens et al. (1990) on the commercialization of new technology finds that successful enterprises take commercialization as the primary goal of innovation. Moreover, they take various overlapping stages into account in the process control (shown in Figure 5-5). Most companies with poor performance tend to equate commercialization with R&D and think that performance is positively proportional to the R&D input.

Wood & Brown (1998) claim that even if the new structure and method have been confirmed in the research environment, they still need further extraction in order to become the technology which can be produced repeatedly, tested, and documented before they meet mass production. The innovation process for the new product can be divided into three stages (shown in Figure 5-6). The two stages of implementation and manufacturing make up the process of technology integration. In Sony's research on the diode, the sample manufacturing equipment is a flexible tool welded manually by researchers; and industrial production needs equipment with quality assurance as well as repeatable production. In addition, because diode was a new technology then, there were no special equipment and tool manufacturers. To develop this technology, production engineering had to be implemented between engineering development and production. American companies took more advantage of external technical forces to complete equipment and tool development manufacturing, but Sony made use of its own capability for tools and equipment development.

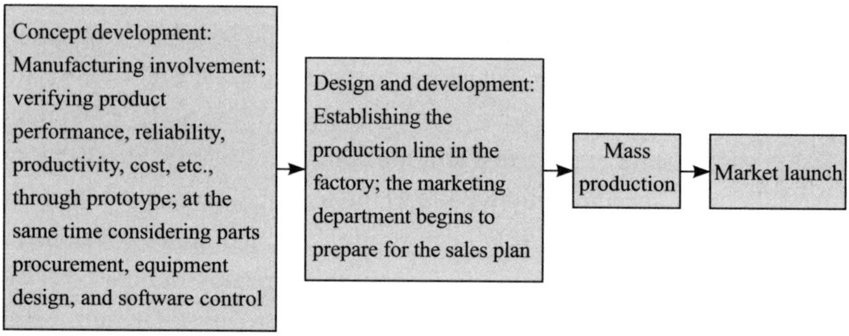

Figure 5-5 The mode of successful enterprises in technology commercialization

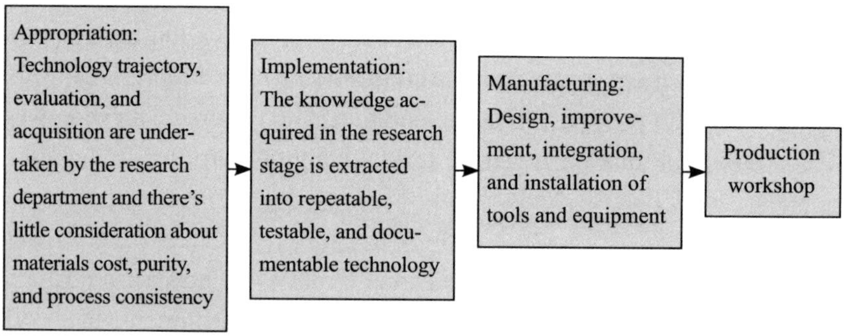

Figure 5-6 Sony's commercialization process of nascent technology

Harryson (2004) deems that Japanese companies have five kinds of mechanisms and principles which guarantee the seamless transfer and convergence from R&D to production: i) Staff accept on-the-job training and strategic rotation, in order to ensure that each member understands the technical factors and the manufacturing requirements in the development process; ii) the researchers move to the production workshop with the products, so as to integrate technical ability into the products; iii) design and manufacturing receive the same attention; R&D is advocated for design and manufacturing, rather than for manufacturing only; iv) respecting the production site; v) in the early period of project R&D, product samples determine whether resources could be inputted as an important basis; developers must consider its manufacturability, and cooperate with the production workshop (shown in Figure 5-7).

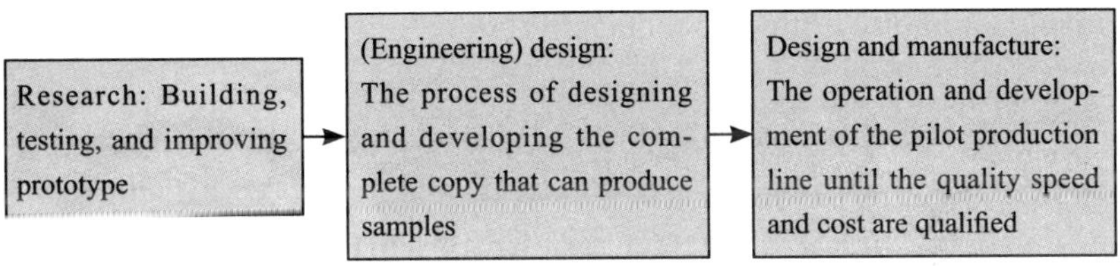

Figure 5-7 Technology integration in technological innovation of Japanese enterprises

Andrew & Sirkin (2003) points out that enterprises have three ways to realize the commercialization of new products: i) Enterprises and the integrators finish the whole task by themselves; ii) enterprises and the orchestrators focus on some stages of commercialization and give the rest to their partner; iii) enterprises and the licensors sell or license product technology to other organizations in order to achieve commercialization. The three ways have their own applicability and shortcomings. The integrators may succeed when they have some skills in product design, in the development of manufacturing technology, in market and cross-functional cooperation. The success of the orchestrators depends on their relationship network. And the licensors distribute more in the biotechnology and information technology fields with rapid changes and high risk.

(2) The importance of manufacturing technology in the process of innovation

First, the manufacturability of the product design. After finishing the development of a product, the mass manufacturing stage comes. If the product does not match the enterprises' manufacturing ability or the environment, or the enterprises cannot carry out mass production on time, the enterprises will suffer loss. P. S. Adler (1995) defines "manufacturability" as the suitability of the product design to the production process. Dean & Susman (1989) think that the design department should adopt various commercial software so as to be able to assess whether the design can be "manufactured." Moreover, enterprises should adopt standard modules as much as possible and use the existing manufacturing ability to assemble, which may improve the manufacturability of new products.

Wheelwright & Clark (1992) regard "manufacturability design" as a method and tool of new product development. Manufacturability design can at first find potential problems by making all sorts of design rules; an important principle is designing for production. Some design rules may require reducing the number of components as much as possible. But if an enterprise blindly reduces the amount of parts and designs simple parts into a complex one, it can lead to the complexity of mold making, an extension of development time, and even the interruption of the process. So they must think about everything carefully when making design rules. In a word, enterprises should fully consider the limits and characteristics of manufacturing system. With the method of empirical analysis, Swink (1999) studies the influence factors of new product manufacturability (NPM). He comes to the conclusion that NPM is related to the complexity of project; the participation of manufacturing and development can reduce the negative influence of the uncertainties of technology.

Second, the development of the manufacturing process. Mansfield (1988) divides enterprise innovation into six stages from applied research, if any, to initial commercialization. i) Applied research: with business purposes, enterprises discover the application of new scientific knowledge to specific products and processes; ii) project requirements and basic specifications: product features, the participation of marketing and other non-R&D personnel, and the project plan making; iii) prototype and trial manufacture: design, prototype test, and plant test; iv) tooling, manufacturing equipment and facilities: production preparation, production facilities, tooling and equipment design, construction, and purchase; v) mass production: worker training, production facilities debugging, and production system commissioning; vi) market start-up: market research, advertisement, promotion, and others. From applied research to the product or the initial industrial commercialization, Japanese enterprises spend 25% less time and cost 50% less than American enterprises. Considering the stage distribution of resources, American enterprises pay more attention to starting the market, while Japanese ones pay more attention to the process of technology development and manufacturing efficiency; 44% of the development cost is used for tooling, equipment, facilities, etc.

After the study on the pharmaceutical industry, Pisano & Wheelwright (1995) find that there are some problems which change the competitive nature of high-technology enterprises, such as the shortening of the life cycle of the product, the increasing design difficulty of the product manufacturing process, the segmentation of the market and more and more similar technologies. Integrating the development of the manufacturing process into the product development is the key to speeding up new product launch, to finishing the trial production quickly, to strengthening the product function, and to improving customer recognition. This can raise the difficulty of entrance of the manufacturing technology that is protected by patents or difficult to be replicated and expand its proprietary positions. In the process of the development of a new product, many problems regarding technology transfer cannot be solved by communication for the enterprises in the pharmaceutical industry. An effective method for testing process design is trial production. The speed of project development depends on the early performance of the project. Taking the development of drugs based on chemistry as an example, the higher the proportion of project resources allocated in the manufacturing process development in the early period is, the shorter the development time is. But this principle does not exist in the development project based on biotechnology.

Pisano (1996) thinks that the connection between product and manufacturing technology is very different from industry to industry. In the assembled product industry, such as the automobile, the computer, consumer electronics, etc., process design has little effect on the function and performance of the product, but in industries such as the pharmacy, the chemical engineering, the semiconductor, and the advanced materials industries, the product design and the process design depend on each other; the change in the process has a great influence on product performance; any tiny change in the product design may require the development of a brand new technology. In these industries, process development plays an integrating role throughout the product development. A major problem of product development is how to transfer new process technology to the workshop. With a survey on a development project in the pharmaceutical industry, Pisano finds that lots of knowledge has been coded and can be simulated by computers. And in addition to technology, it is also necessary to accumulate knowledge regarding organization. Therefore, trajectory selection and amplification is the key step in process development, and a lot of problems can only be solved in the lab or in the pilot trial stage. As for the biological pharmaceutical industry, due to little basic research, practical accumulation, and scale-up production knowledge of bioengineering, it depends seriously on "learning-by-doing" in the workshop.

Xu Qingrui et al. (2000) believe that the ability for an enterprise to timely turn product design into a product with high quality and at a low cost is a question of core abilities. Sun Minggui (2002) thinks that from the perspective of enterprises, technology includes product development technology and manufacturing technology. The former refers to the ability to put forward the conception and create the new product, and the latter refers to the ability to manufacture the product with excellent quality and at a low price based on the conception. The interdependence of the two abilities is the key for enterprises to win in the market competition.

(3) The commercialization of new technology and R&D and the transfer of S&T achievements

There are two kinds of understanding of new technology in the academic circle. One is wide sense understanding, which sustains that the commercialization of new technology includes the whole process from the R&D of the new technology in the laboratory to the launch of the product on the market; this is what Wood & Brown (1998), Nevens et al. (1990), Jolly (2001) and others believe in. The other one is narrow sense understanding, which sustains that the commercialization of new technology mainly refers to the process of technology transfer and profit creation included in the formation of the product and technology, which is what Yang Renfei & Tong Yunhuan (2004) believe in. Only the commercialization of wide sense technology includes technology integration (shown in Figure 5-8).

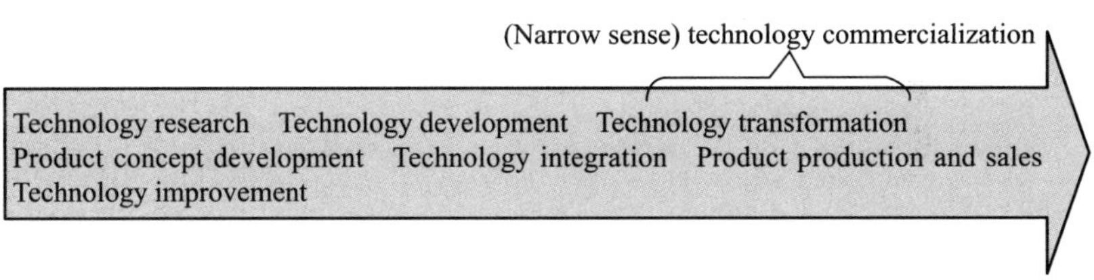

Figure 5-8 (Wide sense) new technology commercialization and integration

It can be seen from Figure 5-8 that the starting point of (wide sense) commercialization of new technologies is the innovation of new technology. The starting point of technology integration is how to apply the prototype to mass production after technology conception has turned into product prototype. Since different achievements developed by the same technology can be applied to different products, in the commercialization of new technology there may be development and integration of a variety of products. Excellent enterprises tend to turn their own R&D capability into products according to the market, and these enterprises often do much better than their competitors in launching their products onto the market rapidly.

5.3 Technology integration from two perspectives

5.3.1 Technology integration from the perspective of the construction of enterprise knowledge

(1) The classification of knowledge

Polanyi (1962) divides knowledge into explicit knowledge and tacit knowledge. Explicit knowledge refers to the systematic language which can be formally coded. It could be combined, stored, and found again, and could be transformed into related cases by different mechanisms. Tacit knowledge refers to the knowledge that is deeply rooted in the individual and his/her mental mode. It is difficult to code and communicate and is obtained mostly based on experience. Tacit knowledge broadens only under specific conditions and in specific domains through relative activities. It can also be obtained through people's observation, imitation, and practice or broadened through "apprentice" training and face-to-face interaction mode, or transferred between organizations with the flow of people.

According to the subject and its form of existence, Collins and Blackler divide tacit knowledge into four kinds: Embodied knowledge, which exists in individuals in the form of personal skills; embrained knowledge, which exists in individuals in the form of personal cognitive ability; embedded knowledge, which exists in organizations in the form of organizational practice; encultured knowledge, which exists in organizations in the form of organizational concepts (Tang Chaoying et al., 2004).

OECD (2000) categorizes knowledge into four types: Know-what, know-why, know-how, and know-who. Among them, the latter two kinds of knowledge, difficult to symbolize and encode, can be called tacit knowledge. Lam (2004) divides enterprise knowledge into four types (shown in Table 5-2); the innovation features of enterprises with a different knowledge type are different.

Table 5-2 Lam's classification of enterprise knowledge

	Individual	Collectivity
Explicit	Embrained knowledge	Encoded knowledge
Implicit	Embodied knowledge	Embedded knowledge

(2) The construction of enterprise knowledge

Iansiti (1995) thinks that the new product is the result of the integration of new knowledge and existing knowledge, and various technologies can add value only if they are integrated into a system (1998). Iansiti studies technology integration of products from the perspective of knowledge integration, namely the construction of systematic knowledge. He divides enterprise knowledge into the knowledge of various disciplines (explicit knowledge) and the knowledge related to a specific environment (mainly tacit knowledge, including the knowledge of enterprise product structure, system design, and the relative manufacturing process as well as the ability of suppliers) (Iansiti, 1995). Systematic knowledge is the interconnection and combination between new technology and the concrete application environment (including the production process, the user environment, the supplier, and the product structure) (Iansiti, 1995). Technology integration is the process of the generation, the application, and the accumulation of systematic knowledge. The technology integration mechanism has three dimensions: Knowledge generation, knowledge accumulation, and knowledge application (Iansiti, 1995, 1998; Iansiti & West, 1997, 1999).

Iansiti carries out a series of research works regarding the semiconductor, the computer, and the software industries in America and Japan, and draws the following conclusions: First, faced with the complex and ever-changing technological environment, American enterprises adopt the method of system focusing. They integrate system knowledge through special integration groups. They not only learn from product renewal, but also from the influence of new technology on the organizational capacity for complex production. Valuable knowledge spreads throughout the process; at the same time, information can have upstream feedback to improve product development performance (Iansiti, 1993, 1997a). Second, in industries with rapid and discontinuous technology development, project performance and technology integration are closely related, and the investment in human resources as well as the time needed for development has a negative correlation with the degree of system focusing. Cross-functional integration is just a kind of organizational mode of integration, and the cultivation of T-type talents is an effective integration method. On the other hand, T-type talents are also good integrators in enterprises (Iansiti, 1995). Third, technology output and integration of products are related positively, but potential and integration of technology have no significant correlation (Iansiti & West, 1997).

Fleck (1994) believes that successful implementation needs the combination of general technical knowledge and enterprise practice knowledge, and that the latter originates in the daily practice of enterprises, which is the key factor of the success of innovation. Tang (1998) claims that the integration of the individual and his/her knowledge and skills are the foundation of enterprise innovation. Grant (1996) thinks that in the process of the importation and application of external advanced technology, it is extremely important that the key competitive advantage of enterprises should come from the effect of technological integration, rather than from the imported technology itself. For an enterprise with imported technology, it is far from enough just to buy technology or capacity and leave them statically. In order to create value from the imported technology, enterprises also need to train workers in using the technology after the transaction has been completed and integrate it into the enterprises' existing resources. Most enterprises need to integrate the professional knowledge which exists in each individual, so they must establish a corresponding integration mechanism to promote the combined application of each individual's professional knowledge. Wang Dazhou (2001) finds that the essence of enterprise technological innovation is the transformation from knowledge to the artificial product or service and the generation of new knowledge.

Nonaka & Takeuchi (1995) regard innovation as a process to find a problem, define it, and create new knowledge and solve the organizations' problem, as well as the organizations' knowledge generation process. And they call it the knowledge production spiral, including four specific steps: i) The socialization (the communication of individual tacit knowledge), which means sharing experience by working together, obtaining process skills by the non-language approach, and then forming resonance; ii) externalization (the transformation from tacit knowledge to explicit knowledge), namely spreading knowledge through metaphor and analogy, and forming conceptual knowledge; iii) combination (of explicit knowledge), which means combining, editing, and processing the existing knowledge with a different carrier to form new knowledge; iv) internalization (the transformation from explicit to tacit knowledge), which means an individual shares ideas and skills through the first three steps to form operational knowledge, which often relies on "learning by doing."

Sherman et al. (2005) analyze the accumulation and exchange of intra-organizational knowledge from three dimensions of the recording, spread, and use of knowledge. Based on the research on high technology enterprises in the US and northern Europe, they find that communication and accumulation between the R&D and marketing departments have a significant effect on product development performance. Zhu Fangwei et al. (2004) elaborate on the transformation of tacit knowledge into technical achievements in the process of technology transfer. The tacit knowledge in their opinion means a certain conception, design, trial manufacture, and "know-how knowledge" in the production process, which cannot be or are not expressed by language, as well as the cultural knowledge and the specific management knowledge, such as the concept, spirit, values of technology achievement of R&D institutions and their staff. Zhu Fangwei divides tacit knowledge into three categories: i) The conception, design, trial production related to some specific technology achievements of R&D personnel and the tacit knowledge of the process of thinking and understanding redesign and reproduction (K1); ii) the operation skills of R&D and production personnel in the stage of specific design, testing, and trial production (K2); iii) the cultural knowledge of R&D institutions and their staff, such as their concept, spirit, and value (K3). He puts forward a kind of "three-stage model," by which the technology acceptance side can explore such tacit knowledge.

5.3.2 Technology integration from the perspectives of internal and external resources

Technology integration involves many kinds of technical resources, and enterprises can choose internal or external resources.

Most enterprises should be the "subject of technology integration" in the innovation network, which means they launch prototype technology into industrial production through the integration of various external resources. Calantone et al. (2002) claim that the key factor from innovation to mass production is the integration of functional resources. Through their research on medical instrument manufacturing enterprises, Parthasarthy & Hammond (2002) find that only increasing investment in R&D cannot significantly increase the probability of the success of enterprise innovation, while the integration of internal functions and external resources play a significant role in the success of innovation; if enterprises can enhance the integration of internal functions when they are increasing investment in R&D, it will have more significant effects on the success of innovation.

(1) The integration of internal resources

First, the way and mode of the integration of internal functions. Wheelwright & Clark (1992) study the integration of different functional resources and raise four integration modes from the two dimensions of the information communication method and the work connection method between upstream and downstream in the process of developing new products: i) Serial/batch transfer mode; ii) (downstream) independent implementation ahead; iii) (downstream) intervention ahead; iv) solving problems together.

Kahn divides the integration of enterprise internal resources into interaction (1996, 2001), collaboration (1996, 2001), and coordination (2001). Among them, interaction refers to the structured and formal communication among different departments, such as project meetings, technical documentation, information management, etc.; collaboration refers to the unstructured relationship among departments, such as team work, mutual understanding, common vision and goals, resource sharing, etc.; coordination refers to information sharing, resource sharing, customer value created by all functions, etc. Gomes et al. (2003)

think that the relationship among different functions includes interface, cooperation, coordination, collaboration, communication, cross- or multi-functional teams, etc.

Dean & Susman (1989) put forward four kinds of mechanisms and what they lack in order to achieve a better combination of design and manufacturing: i) The denial system of manufacturing; its disadvantage is little daily communication of design and manufacturing. ii) Integrators, who must seek a balance point in the design and manufacturing personnel; its disadvantage is that a single person has a big influence on decisions concerning the project process, while the manufacturing personnel cannot start to work before completing the design. iii) The cross-functional team; the first scenario is the participation of the representatives of manufacturing, engineering, quality control, testing, the buyer, and the marketing on the design team, so as to finish the whole process. However, sometimes the design department might think that the requirements of the manufacturing department are not realistic; the second scenario is adopting a team or "product center," which is made up of representatives from the design, manufacturing, quality control, and auxiliary production departments. iv) An independent department in charge of product and process; the first scenario is that independent product and technology branches are set up led by department managers; the second scenario is that the product developers and process developers make up an independent department; the third scenario is that specialized product-process engineers make up an independent department; however, its disadvantage is that it is easy to accept a compromised solution due to the close relationship between the designers and the manufacturing personnel. Dean & Susman (1989) think that the more advanced the form is, the more general the cooperation and the concurrent engineering are, and the higher the requirements of personal skills are. On the contrary, the organizations with relatively stable product and process and limited adaptability to changes should adopt the denial system of manufacturing or integrators.

Nihtilä (1996) sums up four methods of R&D and production in the early stages (product concept and planning phase) of product development through literature and case study: i) The integration of a non-personal method, including standards, procedures, plans, etc.; ii) milestone and design review; iii) personal integrators (a specialist responsible for integration); iv) formal cross-functional teams.

Iansiti's study finds that American enterprises dealing in semiconductors, computers, and other products set up an integration group at the beginning of R&D made up of managers, scientists, and engineers in order to improve the manufacturability of the application and the commercialization of the new technology; this group is responsible for managing and coordinating the entire product development and equipment selection. By investigating the different technology selection schemes of product design and manufacturing process, its main purpose is to find a balance between laboratory research and the capability of the manufacturing system. The characteristics of the integration group are that the members possess the T-type skills structure, they are close to manufacturers; they engage in the production line and relative product development (product flows); and thus they accumulate rich systematic knowledge (Iansiti, 1993, 1995, 1997).

The cross-functional team is not the best organizational mode for all projects. Wheelwright & Clark (1992) claim that when the product development cycle is long, the alteration of product design is slight, the customer requirements are clearly defined, and the communication among the functions is smooth, enterprises can integrate the tasks through procedures and meetings and do not need to adopt the cross-functional integration mode. Sobek II et al. (1998) think that the weaknesses of a cross-functional team are: i) The internal knowledge of the functional department cannot develop in depth. ii) Organizational learning across projects has also dropped because of frequent personnel rotation; iii) Standardization across products has suffered because product teams have become autonomous; iv) Engineers are often torn between the orders of their functional bosses on the one hand and the demands of project leaders on the other. Compared to the development of the cross-functional team of American car enterprises, Toyota is more inclined to the organizational mode of functional departments (shown in Table 5-3). Typically, Toyota's product integration mode builds on its stable long-term employment system and the technical features of the Japanese auto industry.

Table 5-3 The comparison of the organization of product development between Toyota and American auto enterprises

	Toyota mode	American auto enterprises
Organization	Personnel from an assigned functional department to participate in the development without a specific project team	Cross-functional teams
Coordination	Brief meetings when necessary with clear rules about written material and meetings based on written coordination	
Mentoring supervisors	Taught by experts of internal functional departments, deepening professional skills	Learning from cross-functional departments and mastering other skills
Integrated leaders	The chief engineer	The project manager
Skills	Emphasizing standard skills, staff skills training, and work shift on the same functional position	Relying on universities and training institutions to train employees and using work shift on different functional positions
Flexible work standards	Making most development works standardized, and paying attention to the specifics of the project	
Design standards	Using a large number of design standards to guide the design, but paying attention to the improvement and update of the design criteria, which reflect the enterprise production capacity and knowledge accumulation	Tendency of not adopting design criteria

Second, the effect of internal function integration on innovation performance. Through the empirical study on American electronics enterprises, Kahn finds that the collaboration among the marketing, the R&D, and the manufacturing departments has a significant effect on improving product development performance, but the interaction among these departments has no significant effect on product development performance (Kahn, 1996); through the empirical study of the clothing and textile manufacturing industry, Kahn finds that the integration of the three internal functions and the market orientation are important to the development performance of new products (Kahn, 2001). Swink (1999) claims that the participation of the manufacturing department in product development has a positive neutralization effect on the uncertainty of technology.

Gomes et al. (2003) study the relationship between the integration of the R&D and marketing functions and the product development performance of British and Dutch enterprises, and find that the interaction between the R&D and marketing functions has a significant impact on reducing the time required for development and improving the quality of the product. However, a significant effect of this coordination on development performance cannot be confirmed. From the perspective of innovation process, strengthening the interaction of product development and commercialization has certain effects on reducing the time required for development, and the interaction in the stage of conceptualization is beneficial to reducing the cost of development and to improving the quality of the product. In addition, the interaction in the stage of project planning is beneficial to improving the quality of the product. Gomes et al. think that the higher the degree of product innovation is, the more interaction should be strengthened between these two functions.

Song et al. (1998) propose the integration mode with a combination of cross-stage and cross-function; they also divide enterprise innovation into five stages of market opportunity analysis, planning, development, testing, and going public. They think that enterprises should adopt the mode of integrating certain functions in certain stages, rather than integrating all functions from the beginning to the end. In some stages, such as the stage of market planning analysis, development, and testing, integrating the three functions at the same time goes against project performance. In the stage of market opportunity analysis, enterprises should integrate the R&D and marketing functions, but it has a negative effect on innovation performance when enterprises integrate the manufacturing and R&D functions; in the planning stage, enterprises should integrate R&D and manufacturing functions as well as marketing and manufacturing functions; in the development stage, the integration of R&D and marketing and the integration of R&D and manufacturing are good for innovation performance, but the integration of marketing and manufacturing is not beneficial to improving innovation performance; in the test stage, enterprises should integrate R&D and marketing as well as marketing and manufacturing; in the stage of product launching, enterprises should integrate R&D and manufacturing, but the integration of marketing and manufacturing is harmful to innovation performance.

Sherman et al. (2005) discuss the interaction of two departments of the previous projects from the perspective of three dimensions of recording, spread, and use of knowledge. Through the study of high technology enterprises in the United States and northern Europe, they find that the integration of R&D department and previous project knowledge, and the integration of R&D and marketing departments have an effect on product development performance. In addition, Nevens et al. (1990) think that the intervention of manufacturing in the development stage is good for the improvement of the manufacturability of new products.

Third, the way to promote functional integration. Bernasco et al. (1999) study the Texas Instruments and find that interaction can be improved through meetings, committees, telephone communication, and other ways, and that collaboration can be improved through goal sharing, mutual understanding, and other informal ways. Calantone et al. (2002) study the integration of marketing and manufacturing functions in product development in the auto industry, and find that the understanding of manufacturing by the market department is good for communication between and integration of the two departments. The communication methods adopted by researchers and developers at Sony are: i) Research achievements are displayed regularly inside the enterprise; ii) the interpersonal communication network (including internal symposium, etc.); iii) work shift of managers and engineers; iv) transferring knowledge from researchers to developers (Wood & Brown, 1998).

(2) The integration of external resources

First, the integration of supplier resources. Ragatz et al. (2002) find that the cost of raw materials purchase of American manufacturers is more than 50% of that of products. Zhou Linyang (2001) think that the cost of raw materials for new products is almost more than 50% of the sale price of the products. So the supplier has a great influence on the cost, quality, technology, speed, and responsiveness of downstream enterprises. Effectively integrating the suppliers into the supply chain or value chain becomes the key factor for the competition among manufacturers. Peterson et al. (2003) confirm that the uncertainty of technology will improve the willingness to share information, and that information sharing is positively related to development performance; the study also confirms that integrating the suppliers into the process can improve the performance of product development. Primo & Amundson (2002) find that the supplier's ability to carry out quality control is good for its participation in the product development project; the supplier's participation has a significant influence on the quality of the product, but not on the speed and cost of development. By analyzing the relationship between suppliers and development enterprises in the process of new product development and researching the role of suppliers in the supply chain of the new product development, it can be found that the suppliers' intervention in projects is important for shortening the project development cycle, improving the content and quality of the technology of the new product, and reducing the cost of development.

In the process of new product development, the specifications concerning the performance and quality of the products and the raw materials make up an important part of the basic technology and an important method for studying the effective cooperation between the enterprise of the new product development and the supplier of the raw materials. Generally, there are two technical problems in the technical document of new product development projects: i) At the start of the project, due to the vague understanding of the key technical parameters and data related to the new product project, suppliers cannot accurately provide the raw materials which truly conform to the requirements for the development of the new product; ii) due to the limited understanding of the new product and the misrepresentation of the technical data of the new product, the development of the new product and the selection of the raw materials cannot be carried out, thus plenty of time must be spent to modify the technical specifications (Zhou Linyang, 2001).

How to integrate supplier resources effectively is the key point of the related research. Hartley et al. (1997) think that more attention should be paid to the selection of suppliers first. Their study on the assembling industry shows that whether the suppliers can finish their task on time has a significant effect on the schedule of the whole project, and a supplier's technical strength will reduce its delay. Ragatz et al. (2002) think that enterprises should enhance the understanding of the suppliers' ability, share technology and information with the key suppliers, and incorporate the suppliers into the development teams. Peterson et al. (2003) think that an understanding of the suppliers is good for information sharing and integration on both sides; suppliers' participation in the design is an effective method of integration. Primo & Amundson (2002) find that suppliers' ability to carry out quality control is good for their participation in the development of new products.

The study carried out by Von Corswant & Tunälv (2002) of a Swedish auto manufacturer and five of its most important system suppliers shows that the suppliers' internal organization of product development and production and their cooperation with other manufacturers and suppliers are of crucial importance. The close connection of the development team and the product department of the system supplier can improve the manufacturability of the new product design and thus improve the quality and reduce the cost. Ragatz et al. (1997) divide the key factors of integrating suppliers into two categories: The structural relationship and the resource allocation. The structural relationship includes the attention of top managers, the sharing of education and training, the consensus on performance indicators, the trust in the suppliers' ability, formal risk sharing and achievement sharing, formal trust, and the attention of the top management of suppliers; the resource allocation includes intellectual resources (technical information, customer requirements, direct cross-functional and cross-company communication), human resources (joint working, suppliers' participation in the project), and physical resources (united information system, technology sharing, workshop and equipment sharing). Wynstra et al. (2003) underline five aspects of the management of suppliers in the development process: Sorting, mobilization, coordination, speed control, and information sharing.

Wasti & Liker (1999) divide suppliers' participation in design into three aspects: i) Suppliers' influence on decisions concerning the design; ii) the broad range of the technical specifications provided by the manufacturers; iii) the degree of communication. Wasti & Liker study the degree of the common participation of American and Japanese auto enterprises and their suppliers in parts design, and find that in both American and Japanese enterprises, the higher the technical uncertainty is, the deeper the degree of the suppliers' participation in the design is. The technical ability of the suppliers is directly proportional to the degree of their participation in the manufacturers' new product development. Kamath & Liker (1994) study the relationship between Japanese auto manufacturers and their suppliers, and find that Japanese auto manufacturers (clients) divide their suppliers into four categories according to the differences in their technical ability: Partners, mature suppliers, subsidiary suppliers, and other contract suppliers; different ways to carry out integration should be adopted according to different types of suppliers.

Second, the effect of external resources on innovation. Von Hippel (1988) finds that in some industries, such as the scientific instrument equipment and semiconductor process equipment manufacturing industries, almost all the major product innovation or new technology is developed by the users. And in other fields, the suppliers or wholesalers are the main innovators. Kessler (2000) finds that compared with totally depending on internal thought and technology, making full use of external technology and suppliers' information can help reduce the cost of development. Chesbrough (2003) comes up with the "open innovation" concept, which sustains that the enterprises are not only confined to internal R&D, but they also focusing on a combined system that links external and internal technology. This system can allow enterprises to obtain technology from the outside through technology licensing, and also activate some abandoned technology in a closed environment.

Relative to American enterprises, whose technological innovations are based on internal technology, Japanese enterprises rely more on external technology whose innovation performance of external technology is superior to that of internal technology (Mansfield, 1988). In the research on the technology development mechanism of Japanese enterprises, Yu Saito (1996) finds that the cycle from R&D to commercialization of Japanese enterprises is shorter than that of European and American enterprises; the reason is that large Japanese enterprises only exploit the core part of new technology, and give the improvement task of the supporting technology to the related SMEs. SMEs often take advantage of the technology from public research institutes, universities, and other external institutions to make up for the inadequacy of their R&D function in the development of new products. When the marketing function is imperfect, the OEM mode is adopted to produce and supply in order to seek economic efficiency of the function. Harryson's (2004) study on Japanese enterprises shows that sometimes finding the right partners (suppliers) is even more important than grasping technical know-how. Japanese enterprises are good at obtaining technology via the external channel, and with the addition of their internal management of innovations as well as coordination ability, they can quickly achieve the commercialization of new technology. Furthermore, Japanese suppliers have high engineering technology ability and close cooperation with the assembly plants. All of these allow Japanese enterprises to get a larger quantity of new parts in a new product than American enterprises.

Parker & Anderson's (2002) study on HP shows that HP outsources its main design and manufacture tasks to professional suppliers in the process of product innovation, and focuses on product integration, product development coordination, marketing, production, and the supply chain. Besides, HP sets up the position of "supply chain integrator" undertaken by employees who have comprehensive skills; from product concept to product sale, they are in charge of coordinating product development, marketing, production, and logistics (shown in Figure 5-9). This measure improves the efficiency of HP's development of new products.

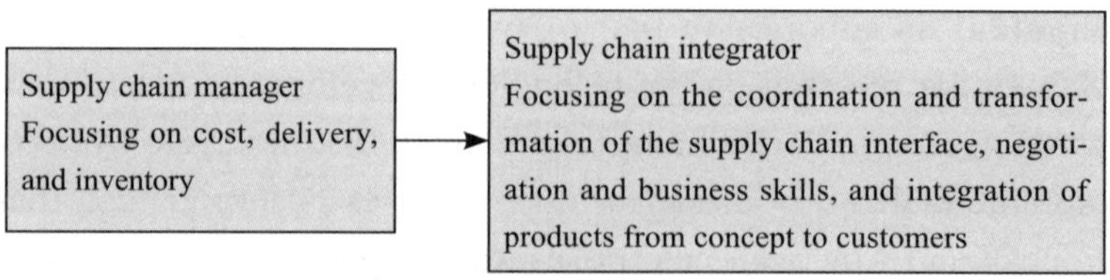

Figure 5-9 The manager and integrator of the supply chain (according to Parker & Anderson, 2002)

Chen Song & Feng Guoan's study (2005) shows that enterprises can no longer solely rely on their internal technical ability to carry out innovation, and the integration of technical resources has become increasingly important. Amir-Aslani & Negassi's study (2006) shows that biotech and pharmaceutical enterprises are faced with increasing competition; they carry out technology integration through the modes of merger and acquisition, R&D alliance, and protocol. Because the R&D cycle is long, and the failure rate is high, more and more pharmaceutical enterprises outsource the R&D to professional companies to speed up the process of new product development. Moreover, many pharmaceutical enterprises tend to trade technology in the late period of R&D. Integration can bring satisfactory efficiency and products. The case study of Hillebrand & Biemans (2004) shows that there are four different relational modes of internal and external cooperation in the process of new product development: i) The mechanism of internal cooperation matching up with external cooperation; ii) the same specifications of internal and external cooperation; iii) external cooperation promoting internal cooperation; iv) internal cooperation is very important for learning from outside partners.

5.4 The feasible mechanism of enterprise technology integration

5.4.1 The participants in and the task boundary of enterprise technology integration

(1) The participants in enterprise technology integration

There are four kinds of participants in technology integration in enterprise innovation: Technology integrator, technology supplier, technology user, and technology supporter. The relationship among the participants may be the relationship among the different functional departments of enterprises; it may also be the relationship among the members of the same team, or among enterprises.

First, technology integrator. The technology integrator is a department or team that takes charge of technology integration, which means transferring technology from the laboratory to the production workshop. The task is different for specific enterprises or projects. For example, some enterprises set up a specific team taking charge of technology integration, and some enterprises do this through their R&D department, and other enterprises do it through their functional departments in turn.

Second, technology supplier. The technology supplier, including internal R&D departments, external enterprises, universities and research institutes, is where product technology and production technology begin. In different integration projects, the technical maturity of the technology provided by technology suppliers is different. According to the type of technology, the technology supplier can be divided into product technology supplier and manufacturing technology supplier.

Third, technology user. The technology user is a department which undertakes the industrialization and the mass production of a new product. If an enterprise organizes production by itself, the technology user and the technology integrator belong to the same enterprise. However, there are also some enterprises that adopt the OEM method of integrating external production resources; the technology integrator and the technology user do not belong to the same enterprise.

Fourth, technology supporter. The technology supporter refers to the supporting resources in the technology integration process; it mainly refers to the provider of raw materials (basic materials or components in the electronics industry), equipment, tooling, and molds. The technology supporter can be divided into durable resource supporter and consumable resource supporter. Durable resources are mainly the resources that can be repeatedly used in the subsequent production process of industrialization, including equipment, tooling, molds, etc.; consumable resources are mainly the resources that need to be constantly imported into the subsequent mass production of industrialization, including raw materials, external processing parts, etc.

The main participants in technology integration and their relationships are shown in Figure 5-10.

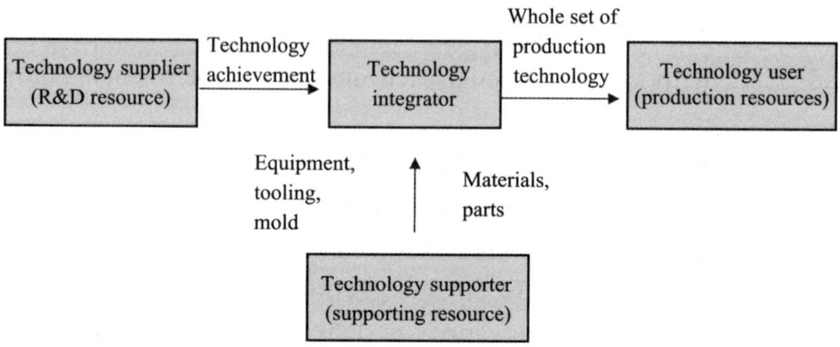

Figure 5-10 The main participants in technology integration and their relationships

(2) The task boundary of enterprise technology integration

Deng Yan (2008) from Tsinghua University finds in her research that even inside an enterprise, different technology integration projects have different technical resource constitutions. For example, according to the specific situation of the project, H Company decides whether to organize the production of a new product by itself or outsource it. When the market prospect of the new product is not clear, it adopts the OEM method in most cases. First, it can reduce risk and allow the enterprise to enter the market quickly; second, the enterprise can get the experience and market feedback of the production organization of the new product from the OEM, and thus the enterprise can improve the product design and produce a second-generation product. Whether taking advantage of internal resources or integrating external resources, different projects of the same enterprise have their own considerations. When the interviewees from the biological pharmaceutical industry introduced their technology sources, they also mentioned that some projects are commercialized based on internal research achievements, and some are commercialized by introducing the technical achievements of external research institutes. Therefore, the technology resources which need to be integrated into the process of technology integration can come from internal or external achievements, whichever is better; it depends on the specific situation of every project. If the resources are divided into internal and external resources according to the boundary of the enterprise, the analysis model cannot be established to apply to all the technical integration projects.

Deng Yan (2008) claims that the division of internal and external resources is based on the boundary of the integration task of enterprise technology, which means defining and analyzing internal and external resources from the perspective of the "technology integrator."

Based on the previous statement, the "external resources" in this book are discussed in relation to the technology integrator, which certainly does not always mean the external resources of enterprises. "Task boundary" and "enterprise boundary" may overlap, which means that enterprises are not engaged in R&D and production, but they integrate external R&D and manufacturing resources as technology integrators. In most cases, "task boundaries" and "enterprise boundary" are different. For example, in Project 1, technology R&D, technology integration, and the production of the new technology are in the same enterprise; the technology integrator integrates the internal R&D and production resources. In Project 2, the technology integrator and the technology supplier belong to the same enterprise; the technology integrator needs to integrate the internal technology resources of the R&D department or the R&D team and external production and manufacture resources. In Project 3, the technology integrator and the producer belong to the same enterprise; the technology integrator needs to integrate the external R&D resources and the internal production and manufacture resources. These situations exist in the 33 technology integration projects investigated by Deng Yan (2008).

In the model of resource integration based on the technology integration task, the subject of technology integration is the department or team undertaking the task of technology integration in the enterprise.

Therefore, in this book, the study of internal technology resources is the organization and management of technology resources of the technology integrator itself (not the internal resource integration of the whole enterprise). Whether technology is from an internal R&D department of enterprises or from an external organization (external enterprise or R&D institute), in this model, all the technologies are regarded as external technology resources. Similarly, whether the producer of a new technology product is an internal production department or an external production organization, the producer is considered to be the "technology user" of the external resources. In this book, regarding the integration of the external resources, the authors focus on how to select, organize, and manage the three kinds of external technology resources—the technology provider, the technology user, and the supporter for the technology integrator.

(3) Three dimensions of enterprise technology integration mechanism

In the process of technology integration, enterprises need to integrate their internal resources, the related experiences of past projects and the related external technology resources. The selection and extraction of external and internal technology resources by the enterprise is a dynamic process. So when people analyze technology resource integration, in addition to considering the organizational dimension and the spatial dimension, the temporal dimension also needs to be considered, which means the management of the process and the control of the technology integration project. Therefore, this book states the technology resource integration of the technology integrator in the aspects of the integration of internal resources (organizational dimension), the integration of external resources (spatial dimension), and the management and control of the project process (temporal dimension).

This book constructs the analysis framework for technology integration on three levels: The first level is the structural layer, which presents the dimensions of technology integration; the second level is the mechanism layer, which establishes the approaches to technology resource integration; the third level is the measures layer, which establishes the specific ways of carrying out technology resource integration. The technology integration mechanism proposed in this book is the way that the technology integrator promotes and implements internal and external resource integration in the process of technology integration. In this book, combined with the analysis of the enterprises that were interviewed, there are six categories of the approaches (technology integration mechanism) to integrating internal and external technology resources (shown in Table 5-4).

Table 5-4 The technology integration mechanism and content of the investigated enterprise

Integration dimension (structural layer)	Integration method (mechanism layer)	The main content (measures layer)
Internal integration	Environmental mechanism	The established regulation, system, culture, and customs, which promote the integration of internal resources
	Organizational mechanism	The way to organize the personnel of technology integration in the project
Interface integration	Communication mechanism	The communication and information exchange mode between technology integrator and other participants
	Coordination mechanism	The way to organize and control the technology integrator with respect to the other participants in the project progress
Process integration	Control mechanism	Establishment and management of the process of the technology integration project
	Exchange mechanism	The exchange of information among the members of the different stages in the technology integration of the project

Deng Yan (2008) finds that both the communication mechanism of interface integration and the exchange mechanism of process integration promote technology resource integration through information communication and exchange, but these two types of mechanisms have different focuses. The communication mechanism of interface integration mainly refers to the information communication between the technology integrator and other participants. While the exchange mechanism of process integration mainly refers to the communication between upstream and downstream sectors which undertake different tasks within the technology integrator. From the view of nature, the approaches of technology resource integration can be divided into formal approaches and informal approaches; each type of integration mechanism can also be divided into structured and unstructured mechanism accordingly. Structured technology integration mechanism is the approach which is realized by formal, institutionalized, or clearly described (tangible) integration way; unstructured integration mechanism is the approach which is realized by informal and

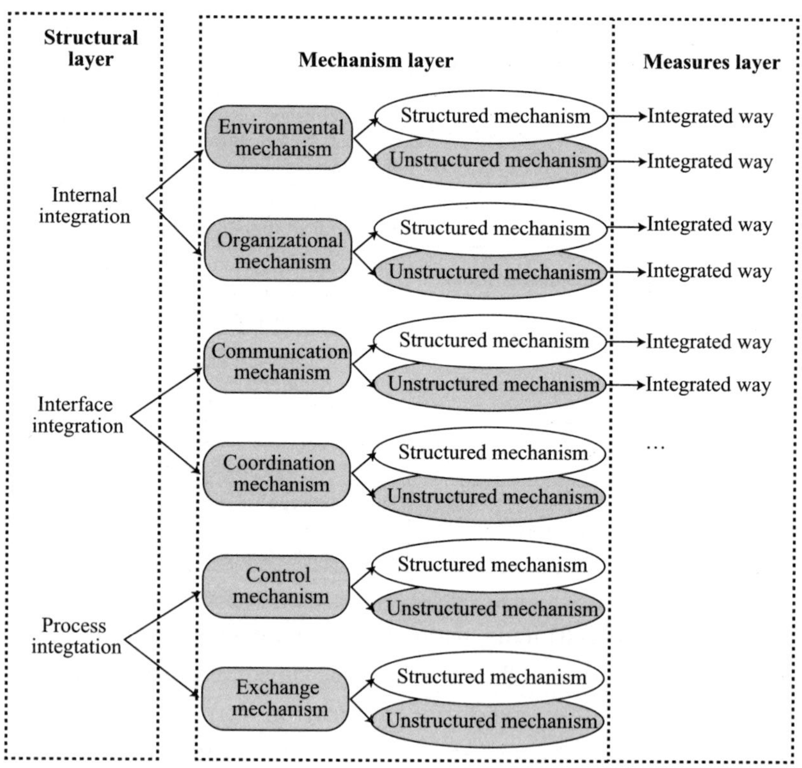

Figure 5-11 The three levels of the technology integration mechanism

difficultly described (intangible) integration way. For example, in the communication mechanism of interface integration, structured communication mechanism mainly refers to communication ways with formal channels, while unstructured communication mechanism mainly refers to informal communication ways. Therefore, enterprises can design and choose at least 12 different types of technology integration mechanisms (shown in Figure 5-11).

5.4.2 The mechanism of the three aspects of enterprise technology integration

(1) The three aspects of the enterprise technology integration mechanism

As can be seen from Figure 5-11, each dimension of technology integration has four kinds of integration mechanisms, and each kind of technology integration mechanism can be implemented in several different ways.

(2) The approach and method of the realization of enterprise technology integration

The implementation of the technology integration project in the process of enterprise innovation can integrate all kinds of technical resources from three dimensions with 12 approaches and 30 integration ways. Although different projects adopt different integration mechanisms and integration ways, there are still some common characteristics. The essence of technology integration is that enterprises integrate various internal and external resources concerning technology. Deng Yan (2008) finds that technology integration has three aspects. The first is the structural layer, which includes three dimensions of technology integration (internal integration, interface integration, and process integration); the second is the mechanism layer, which includes 12 technology integration approaches (namely the technology integration mechanism); the third is the measures layer, which includes 30 specific methods of technology integration (shown in Table 5-5).

Table 5-5 The dimension, mechanism, and realization methods of technology integration

Structural layer (technology integration dimension)	Mechanism layer (technology integration method)	Measures layer (realization methods of technology integration)
Internal integration	Structured environmental mechanism	1. Technical document management and information construction; 2. the use of the design rule and method; 3. regular communication system of different departments
	Unstructured environmental mechanism	4. Communication among technical personnel; 5. not blindly pursuing advanced and practical technology, paying attention to practicality of the new technology[1]; 6. resources and information can be shared in different departments
	Structured organizational mechanism	7. The integration of the project organization structure[2]; 8. the project leader is responsible for the whole process of project implementation; 9. project core members transfer with the project and participate in the whole process of the project; 10. the evaluation of project members is directly tied to the project performance
	Unstructured organizational mechanism	11. Most project members have the experience of similar projects; 12. choosing interdisciplinary technical personnel with comprehensive background to participate in the project[3]
Interface integration	Structured communication mechanism	13. Importing technical documentation; 14. the formal communication in the integration process
	Unstructured communication mechanism	15. Having a complete understanding of the participants before the integration; 16. personnel of both sides should communicate and exchange in time during the integration process

[1] The enterprise culture, which pays attention to technology practicality, has the advantage of making different functional departments share the same goal, and thus it is conducive to the integration of internal resources.

[2] For example, setting up a project team, cross-functional team, technology integration group, centralized management of outsourcing tasks, etc.

[3] Interdisciplinary technical personnel mainly refer to technical staff who, in addition to their professional knowledge, know about research, development, engineering, equipment, procurement, and production. Comprehensive background mainly means their understanding of their departments, as well as the specific circumstances of other departments in the enterprise.

(Continued)

Structural layer (technology integration dimension)	Mechanism layer (technology integration method)	Measures layer (realization methods of technology integration)
Interface integration	Structured coordination mechanism	17. Taking some integrated organization models[1]; 18. the performance of the final industrial production stage has an important influence on the interest allocation of other participants (mainly for the technology provider interface)
	Unstructured coordination mechanism	19. Other participants are involved in the whole project; 20. the technology integrator masters the decision-making right of the technical scheme; 21. other participants are replaceable
Process integration	Structured control mechanism	22. Making a comprehensive evaluation before the approval of the project; 23. establishing the project plan, transferring the standard, and strict guidelines of implementation; 24. establishing the control point during the process, many departments should have a strict audit on the scheme; 25. doing different tasks of the project at the same time
	Unstructured control mechanism	26. The enterprise's top management attaches great importance to the project and coordinates the progress of the project; 27. to fully consider various problems in the earlier stage of industrialization production; 28. the trial production fully simulates the real production environment[2]
	Structured exchange mechanism	29. During each stage, the technical staff can communicate through project meetings and technical documents
	Unstructured exchange mechanism	30. Technology personnel timely communicate and exchange in each stage

Although different projects of technology integration have different methods of integration and integration mechanisms, Deng Yan (2008) also finds out that Chinese enterprises have the following features in the application of the technology integration mechanism and method:

[1] It includes the forms: The importing of external technical personnel, the establishment of a united team, and designating special contacts.

[2] If the trial production has full simulation of the real production environment, people can discover potential problems in the technical solution as soon as possible. The interview finds that it is not the key to arranging the trial production in the process of the project; instead, the key is the degree of the simulation of the production environment.

Considering the structural layer, technology integrators pay different attention to the three dimensions of technology integration. Among them, technology integrators pay the least attention to interface integration and equal attention to internal integration and process integration.

Regarding the mechanism layer, technology integrators put different emphases on each kind of technology integration mechanism. In the same dimension of technological integration, different projects have significant differences in the application of the four kinds of integration mechanisms. In the dimension of internal integration, the project pays the least attention to the structured environmental mechanism, followed by the unstructured organizational mechanism; in the dimension of interface integration, the unstructured communication mechanism is the most common in project application, and the unstructured coordination mechanism has the lowest degree of application; in the dimension of process integration, the unstructured control mechanism is the most common in project application.

Concerning the measures layer, the nature and the scale of the enterprise have a certain effect on the project integration mode. The nature of the enterprise influences the application of technology integration, which mainly concentrates on interface integration. Compared with the enterprises dominated by production, the enterprises dominated by R&D tend to let other participants be involved in technology integration in the whole project; compared with the enterprises dominated both by production and R&D, the enterprises dominated by production tend to focus on the integration of formal channels. And the influence of the scale of the enterprise on the method of technology integration is reflected in all of the three dimensions. The enterprises with fewer than 100 people adopt the following mechanisms less than the enterprises with more than 100 people: The management of technical document and information construction, the application of design rules and methods, the designation of project leaders, the attention and coordination of the top leaders of the technology integration project, and the mastery of the control rights of the technology project.

1) The method of technology integration in most enterprises does not adjust to the changing environment in the process of technology integration. At the same time, technology source has an important influence on the relationship between the technology integration mode and project performance. Therefore, enterprises need to strengthen the comprehensive application ability of the technology integration method and use different modes of technology integration in different technology environments.

The so-called technology environment refers mainly to the various combinations of source and maturity of technology. The source of product technology and manufacturing technology can be divided into three classes, namely internal or cooperative R&D, introduction from external enterprises, and introduction from external research institutes or universities. At the same time, product technology and manufacturing technology can be divided into five grades according to the degree of maturity.

In the survey on technology maturity by Deng Yan (2008), whether it is product technology or manufacturing technology, enterprises tend to select the more mature technology. Regarding the source of technology, surveys show that internal or cooperative R&D is the major source of product technology and manufacturing technology. However, by analyzing the interviews carried out on enterprises, based on the following three cases, it can be found that enterprises regard some research achievements, which are made by research institutes and universities, as their internal R&D. Among them, first, after some research institutes become enterprises, their research achievements are directly commercialized within the enterprise itself; second, the researchers at research institutes or universities join the enterprises along with their technologies which are naturally considered internal research achievements; third, the researchers at research institutes or universities create their own companies and promote the commercialization of their research achievements. Therefore, the role of research institutes and universities in enterprise innovation cannot be judged only from the proportion of technical sources.

Deng Yan (2008) analyzes the influence of the source and maturity of product technology and manufacturing technology on the method of technology integration. She finds that the above four factors have little impact on the method of enterprise technology integration; when the technology environment (mainly referring to technology source and technology maturity) changes, only two technology integration methods change in the project application among all 30 methods. To be specific, compared with the project of product technology introduction from external enterprises, the one from research institutes is more concerned with the introduction of technical documentation. In addition, sharing resources and information among departments has a certain degree of influence on the selection of enterprise technology. Further research on the provider interface of product technology and manufacturing technology has not found that the integration mode of the two interfaces adjusts to the change of the technology environment. Further empirical research has found that the technology source, as the control variable, has a certain influence on the relationship between the integration mechanism (mainly referring to the interface integration mechanism) and project performance. Therefore, in the process of technology integration, enterprises need to strengthen their ability for comprehensive application of the integration mechanism and use different technology integration modes in different technology environments.

As stated earlier, the interface integration dimension has four kinds of integration mechanisms, namely the structured communication mechanism, the unstructured communication mechanism, the structured coordination mechanism, and the unstructured coordination mechanism. Technology integrators need to integrate five different kinds of interfaces, namely the product technology provider, the manufacturing technology provider, the technology user, the raw material supporter, and the equipment supporter. Because of the non-homogeneity of them, a unified management mode and method cannot be adopted for all interfaces during the technology integration process. The enterprise needs to pay more attention to the integration management of the following three kinds of interfaces:

Technology user interface. Enterprises should strengthen the application of the structured communication mechanism, and the specific measures include the communication meeting with production departments and the training for production departments. In addition, strengthening the unstructured coordination mechanism can improve the total performance of technology integration project; there are two integration methods at the measures level: The first is allowing the producer to get involved in the overall process of the project of technology integration; the second is making an alternative offer of productive resources in advance.

Key raw material supporter interface. Enterprises should strengthen the application of the structured communication mechanism, including two kinds of technology integration methods: The first one is to pay attention to the importing of data and document matching technologies, when importing raw materials; the second one is to strengthen the communication and exchange with the key raw material supporter. In addition, during interface integration, the structured coordination mechanism cannot be used too much, such as the joint team, the importing of supplier technical personnel; otherwise, it will affect production efficiency to a certain extent.

Key equipment supporter interface. The enterprise should strengthen the structured coordination mechanism, including the participation of the equipment supporter's technical personnel in the technology integration team, the introduction of technical personnel, the designation of person to contact for the key equipment supporter, etc.

The survey on the actual situation of enterprise technology integration projects finds that the project pays the least attention to the interface integration in the three dimensions, especially the interface integration of the technology supporter. And the result of the performance analysis revealed that the integration mechanism, which has a significant influence on project performance, centers on the interface of technology users and technology supporters. Therefore, technology integrators should raise the amount of attention they pay to interface integration, especially to the interface of technology users and technology supporters.

2) In the internal integration dimension, the unstructured environmental mechanism significantly promotes the technology integration effect.

The related mechanisms of the internal integration dimension include structured environmental mechanism, unstructured environmental mechanism, structured organizational mechanism, and unstructured organizational mechanism. Among them, the unstructured environmental mechanism mainly refers to the mechanism of the cultural atmosphere of the enterprise. Enterprise culture is the concept and the convention, which gradually form in the process of long-term operation and get commonly acknowledged within the enterprise. The formation of enterprise culture needs a long process of accumulation; the system mechanism and the organizational mechanism are more likely to attract the managers' attention than the enterprise culture mechanism. The enterprise culture mechanism can promote project performance more significantly than the former two. Therefore, enterprise managers should pay attention to the construction of enterprise culture, and create an interdependent and pragmatic enterprise culture among departments and technical personnel, instead of relying too much on the rules and regulations and the skills of the organizational structure of the project in order to promote the integration of internal resources. At the implementation level, the enterprise cultural integration mechanism includes encouraging communication among technical personnel, paying attention to technology practice and information sharing among different departments, etc.

3) The structured control mechanism has a significant effect on the performance of the technology integration project in the dimension of process integration.

The mechanisms related to the process integration dimension include the structured control mechanism, the unstructured control mechanism, the structured communication mechanism, and the unstructured communication mechanism. In the dimension of process integration, adopting the structured control mechanism to strengthen the management and control of the entire process will promote and enhance the performance of the technology integration project.

4) The mechanism issues should be comprehensively focused on in enterprise technology integration. Production-oriented technology integration is one of the core issues of enterprise technological innovation, and also one of the main channels for an enterprise to achieve independent innovation. Enterprise innovation should integrate various technology resources regarding three dimensions and via 12 approaches and 30 integration methods.

First, the ability for comprehensive application and for strain in various technology integration methods should be enhanced. There is no big difference in the methods of technology resource integration of different projects in different technological environments (mainly referring to source and maturity of technology). Especially, in the integration of the technology provider interface, most enterprises may not adopt different interface integration modes due to the difference in the technology selected. The results of the performance analysis show that as a control variable, technology source has a certain influence on the relationship between the resource integration mechanism (mainly interface integration mechanism) and project performance. Therefore, enterprises should strengthen their ability for comprehensive application of the integration mechanism. When the technology environment changes, the enterprise should make an adjustment in the application of the integration mechanism, so as to improve its capacity for strain and its adaptation to the changes of the technology environment.

Second, particular attention should be paid to the interface integration dimension; the interface integration management skill should be improved; and the interfaces of different nature should adopt different integration modes. In the three dimensions of technology integration, the attention of the project on the interface integration dimension is obviously less than it is on the other two dimensions. The result of the performance analysis shows that the interface integration mode, especially the integration of the technology integrator on the technology user, the key raw material supporter, and the key equipment supporter, has a significant influence on the performance of the final project. Moreover, the influence of the individual integration mechanism on performance in different interface applications is negative. Therefore, enterprise technology integrators should adopt different integration modes in different natures of interfaces, while paying particular attention to the interface integration dimension.

Third, the project process management should be strengthened by the aid of the control mechanism. In the dimension of process integration, the structural control mechanism plays a positive role in the performance of the technology integration project. Therefore, during the technology integration process, enterprises should strengthen the control and management of the process of the project by the aid of the control mechanism. The project process management can be strengthened in four ways: The first one is by making a comprehensive evaluation on the feasibility of the technology, on the industrial supporting situation, and on market prospects before setting up the project; the second one is by making a project plan and establishing transfer standards, and implementing them strictly; the third one is by setting up periodical control points during the process of project implementation, and organizing more departments to audit the technical solutions strictly; the fourth one is by boosting different project tasks in a parallel way. In the process of technology integration, enterprises should pay attention to the non-structured control mechanism. Especially, enterprises should sufficiently consider various problems in industrialized mass production in the early stages of technology integration.

Fourth, emphasis should be put on building up an enterprise culture and an atmosphere that is beneficial to technology integration. In the internal integration dimension, the non-structured environmental mechanism, namely the enterprise culture beneficial to resource integration, has a positive effect on the performance of the technology integration project. Therefore, enterprise managers should pay particular attention to the construction of enterprise culture, and build a kind of interdependent and pragmatic enterprise culture among departments and technical personnel.

References

Adler, P.S. Interdepartmental interdependence and coordination: The case of the design/manufacturing interface. *Organization Science*, 1995, 6 (2):147-167.

Amir-Aslani, A. & Negassi, S. Is technology integration the solution to biotechnology's low research and development productivity? *Technovation*, 2006, 26 (5): 573-582.

Andrew, J.P. & Sirkin, H.L. Innovating for cash. *Harvard Business*

Review, 2003, 12 (3): 11-16.

Bernasco, W., Weerd-Nederhof, P.C., Tillema, H. & Boer, H. Balanced matrix structure and new product development process at Texas Instruments materials and controls division. *R&D Management*, 1999, 29 (2): 121-131.

Brown, J.S. & Duguid, P. The universities in the Digital Age. *Times Higher Education Supplement*, May 10, 1996: 1-4.

Brown, J.S. & Duguid, P. *The Social Life of Information*. Boston, MA: Harvard University Press, 2000.

Calantone, R.J., Cavusgil, S.T. & Zhao, Y. Learning orientation, firm innovation capability, and firm performance. *Industrial Marketing Management*, 2002, 31: 515-524.

Chen, S. & Feng, G. The technological effectiveness of three technological sourcing modes. *Science Research Management*, 2003, 24 (3): 58-62.

Chen, S. & Feng, G. Functional division of labor in the process of technological innovation. *Science Research Management*, 2005, 26 (1): 25-28.

Chesbrough, H. *Open Innovation: The New Imperative for Creating and Profiting from Technology*. Boston: Harvard Business School Press, 2003.

Dean, J.W. & Susman, G.I. Organizing for manufacturable design. *Harvard Business Review*, 1989, 35 (2): 38-42.

Deng, Y. Technology integration towards industrial production. Doctoral dissertation of Tsinghua University, 2008.

Deng, Y. & Lei, J. From the prototype technology to manufacturing: Technology integration for commercial production. *Science of Science and Management of S.&T.*, 2006 (6): 5-8.

Fleck, J. Learning by trying: The implementation of configurational technology. *Research Policy*, 1994, 23 (6): 637-652.

Fu, J. *Theory on Technological Innovation*. Beijing: Tsinghua University Press, 1998.

Gomes, J.F.S., de Weerd-Nederhof, P.C., Pearson, A.W. & Cunha, M.P. Is more always better? An exploration of the differential effects of functional integration on performance in new product development. *Technovation*, 2003, 23 (3): 185-191.

Grant, R. M. Toward a knowledge-based theory of the firm. *Strategic Management Journal*, 1996, 17 (S2): 109-122.

Harryson, S. *Japanese technology and innovation management: From know-how to know-who*. Hua, H., Li, D. & Hua, H. (trans.) Beijing: Peking University Press, 2004.

Hartley, J.L., Zirger, B.J. & Kamath, R.R. Managing the buyer-supplier interface for on-time performance in product development. *Journal of Operations Management*, 1997, 15 (1): 57-70.

Hillebrand, B. & Biemans, W.G. Links between internal and external cooperation in product development: An exploratory study. *Journal of Product Innovation Management*, 2004, 21 (2): 110-122.

Holatius, K. Cultural adjustment in international technology transfer. *International Journal Technology Management*, 1995, 10 (7-8): 676-686.

Iansiti, M. Real-world R&D: Jumping the product generation gap. *Harvard Business Review,* 1993, 71 (3): 138-147.

Iansiti, M. Technology integration: Managing technological evolution in a complex environment. *Research Policy*, 1995, 24 (4): 521-542.

Iansiti, M. From technological potential to product performance: An empirical analysis. *Research Policy*, 1997, 26 (3): 345-365.

Iansiti, M. *Technology Integration: Making Critical Choices in a Dynamic World*. Boston: Harvard Business School Press, 1998.

Iansiti, M. & West, J. Technology integration: Turning great research into great products. *Harvard Business Review*, 1997, 75 (3): 69-78.

Iansiti, M. & West, J. From physics to function: An empirical study of R&D performance in the semiconductor industry. *Journal of Product Innovation Management*, 1999, 16 (4): 385-399.

Jolly, V.K. *Commercializing New Technologies: Getting from Mind to Market*. Boston: Harvard Business School Press, 2001.

Jonash, R.S. *The Innovation Premium: How Next Generation Companies Are Achieving Peak Performance and Profitability*. New York: The Perseus Books Group, 2001.

Kahn, K.B. Interdepartmental integration: A definition with implications for product development performance. *Product Innovation Management*, 1996, 13 (2): 137-151.

Kahn, K.B. Market orientation, interdepartmental integration, and product development performance. *Product Innovation Management*, 2001, 18 (5): 314-323.

Kamath, R.R. & Liker, J.K. A second look at Japanese product development. *Harvard Business Review*, 1994, 73 (11): 154-170.

Katz, R., Rebentisch, E.S. & Alien, T.J. A study of technology transfer in a multinational cooperative joint venture. *IEEE Transactions on Engineering Management*, 1996, 43 (1): 97-105.

Kessler, E.H. Tightening the belt: Methods for reducing the cost of new product development projects. *Journal of Engineering and Technology Management*, 2000, 17: 59-92.

Kim, L. Technology policies and strategies for developing countries: Lessons from the Korean experience. *Technology Analysis & Strategic Management*, 1998, 10 (3): 311-324.

Lam, A. Societal institutions, learning organizations and innovations in the knowledge economy. In Christensen, J.L. & Lundvall, B.-A. (ed.) *Product Innovation, Interactive Learning and Economic Performance* (*Research on Technological Innovation, Management and Policy*, Volume 8). Bingley, BD: Emerald Group Publishing Limited, 2004: 43-67.

Mansfield, E. Industrial innovation in Japan and the United States. *Science*, 1988, 241 (30): 1769-1774.

Nevens, T.M., Summe, G.L. & Uttal, B. Commercializing technology: What the best companies do. *Harvard Business Review*, 1990, 42 (3): 154-163.

Nihtilä, J. Integration mechanisms in new product development. *Acta Polytechnica Scandinavia, Mathematics Computing and Management in Engineering*. Helsinki, Finland, 1996, 78.

Nonaka, I. & Takeuchi, H. *The Knowledge-creating Company: How Japanese Companies Create the Dynamics of Innovation*. New York: Oxford University Press, 1995.

OECD. Urban and regional development in the learning economy.

OECD Publication, 2000.

Oslo Manual. *Guidelines for Collecting and Interpreting Innovation Data*. Paris: OECD Publishing, 2005.

Parker, G.G. & Anderson, E.G. From buyer to integrator: The transformation of the supply-chain manager in the vertically disintegrating firm. *Production and Operations Management*, 2002, 11 (1): 75-91.

Parthasarthy, R. & Hammond, J. Product innovation input and outcome: Moderating effects of the innovation process. *Journal of Engineering and Technology*, 2002, 19 (1): 75-91.

Peterson, G.D., Carpenter, S.R. & Brock, W.A. Uncertainty and the management of multistate ecosystems: An apparently rationale route to collapse. *Ecology*, 2003 (84): 1406-1411.

Pisano, G.P. Learning-before-doing in the development of new process technology. *Research Policy*, 1996, 25 (7): 1097-1119.

Pisano, G.P. & Wheelwright, S.C. The new logic of high-tech R&D. *Harvard Business Review*, 1995, 67 (2): 142-150.

Polanyi, M. *Personal Knowledge: Towards a Postal-critical Philosophy*. London: Routledge & Kegan Paul Ltd., 1962.

Primo, M.A.M. & Amundson, S.D. An exploratory study of supplier relationships on new product development outcomes. *Journal of Operations Management*, 2002, 20 (1): 33-52.

Ragatz, G.L., Handfield, R.B. & Petersen, K.J. Benefits associated with supplier integration into new product development under conditions of technology uncertainty. *Journal of Business Research*, 2002, 55 (5): 389-400.

Ragatz, G.L., Handfield, R.B. & Scannell, T.V. Success factors for integrating suppliers into new product development. *Journal of Product Innovation Management*, 1997, 14 (3): 190-202.

Robinson, W.T. Product innovation and start-up business market share performance. *Management Science*, 1990, 36 (10): 1279-1289.

Sherman, J.D., Berkowitz, D. & Souder, W.E. New product development performance and the interaction of cross-functional integration and knowledge management. *Journal of Product Innovation*

Management, 2005, 22 (5): 399-411.

Sobek II, D.K., Liker, J. & Ward, A.C. Another look at how Toyota integrates product development. *Harvard Business Review*, 1998, 12 (1): 36-49.

Song, X., Thieme, R.J. & Xie, J. The impact of cross-functional joint involvement across product development stages: An exploratory study. *Journal of Product Innovation Management*, 1998 (15): 289-303.

Sun, M. Analysis of outsourcing strategy of Japan enterprise. *Foreign Economies and Management*, 2002, 3 (7): 45-52.

Swink, M. Threats to new product manufacturability and the effects of development team integration processes. *Journal of Operations Management*, 1999, 17 (6): 691-709.

Tang, C., Zhou, J. & Liu, T. Research on the absorption model of enterprise tacit technical knowledge. *Science Research Management*, 2004, 37 (4): 41-50.

Tang, H. An integrative model of innovation in organizations. *Technovation*, 1998, 18 (5): 297-309.

Von Corswant, F. & Tunälv, C. Coordinating customers and proactive suppliers: A case study of supplier collaboration in product development. *Journal of Engineering and Technology Management*, 2002, 19 (3-4): 249-261.

Von Hippel, E. *The Sources of Innovation*. New York: Oxford University Press, 1988.

Wang, D. The use of knowledge in the process of enterprise technology innovation: Comparison of China and the West. *Scientific Management Research*, 2001, 19 (4): 1-4.

Wasti, S.N. & Liker, J.K. Collaborating with suppliers in product development: A US and Japan comparative study. *IEEE Transactions on Engineering Management*, 1999, 46 (4): 444-461.

Wheelwright, S.C. & Clark, K.B. *Revolutionizing Product Development: Quantum Leaps in Speed, Efficiency and Quality*. New York: Free Press, 1992.

Wood, S.C. & Brown, G.S. Commercializing nascent technology: The case of laser diodes at Sony. *Journal of Product Innovation Management*, 1998, 15 (2): 167-183.

Wynstra, F., Weggeman, M. & Van Weele, A. Exploring purchasing integration in product development. *Industrial Marketing Management*, 2003 (32): 69-83.
Xia, D., Shi, D. & Zhang, L. *Automobile Industry Technical Progress and Industrial Organization*. Shanghai: Shanghai University of Finance and Economics Press, 2002.
Xu, Q., Guo, B. & Wang, Y. Development of Technological Innovation in China: Core competence-based innovation portfolio. *Journal of Industrial Engineering/Engineering Management*, 2000, 14 (S): 1-8.
Yang, R. & Tong, Y. Discrimination in the strategy's benefit of enterprise technology assessment. *Science and Technology Management Research*, 2004 (3): 29-32.
Yu, S. *Technology Development: The Technology Development Mechanism and Policy of Japan*. Beijing: Science and Technology Literature Press, 1996.
Zhou, L. The role of suppliers in new product development and supply chain. *Technology & Economics in Petrochemicals*, 2001, 17 (1): 53-62.
Zhu, F., Tan, Y. & Tang, L. Research on tacit knowledge conversion during technology transfer. *Science of Science and Management of S.&T.*, 2004 (11): 79-82.

Chapter 6

The institutional guarantee of Chinese independent innovation: The construction of an innovation system

6.1 Innovation system, the institutional guarantee of independent innovation

6.1.1 The connotations of innovation system

The British economist Christopher Freeman (1987) explicitly put forward the concept of national innovation system (NIS), when he studied the experience of the Japanese economic take-off. He pointed out that an NIS is "a network of institutions in public and private sectors in which the attitudes and interactions begin, import, modify, and spread new technologies." In 1996, the OECD (Organization for Economic Cooperation and Development) gave a more comprehensive and systematic summary of an NIS—The innovative performance of a country depends to a large extent on how the actors involved in innovation relate to each other as elements of a collective system of knowledge creation and use as well as the technologies they use. Lundvall (1992) regards an innovation system (IS) as elements and relationships which interact in the production, diffusion and use of new, and economically useful knowledge.

The authors in this book believe: First, in certain organizations, an IS is an organic combination of institutional arrangements and mechanisms, which is beneficial to the initiation, implementation, and diffusion of innovation. In other words, an innovation system is a network system in certain range that consists of the innovation-related subjects and mechanisms which promote the production, use, and diffusion of knowledge or technology. Second, a national innovation system is a part of the social subsystems which is beneficial to the initiation, diffusion, and persistence of technological innovation, including a group of innovation subjects, such as enterprises, universities, scientific research institutes, innovation incubators, etc.; a set of innovation mechanisms, such as the division and cohesive mechanism of different innovation subjects and the society credit mechanism; a set of supporting systems, such as the supply system of industrial generic technology, the social culture that encourages innovation, and the intermediary service system that supports innovation. An IS is interdependent, open, and in continuous development. It is made up of a very complex network of interaction between the market and the non-market. Third, there are innovation systems on at least four levels within a country, namely country, industry, region, and enterprise. A sound innovation system has at least two characteristics: One is effectiveness, which means that various subjects and relevant factors that make up the IS must be effective. Enterprises, scientific research institutes, universities, government, and intermediary institutions should have remarkable ability, efficiency, and effect. The other one is coordination. This means that only if the innovation subjects and elements communicate extensively, knowledge or technologies disseminate and spread effectively, and innovation resources are allocated rationally, will the IS be efficient on the whole.

6.1.2 The practices of building up a national innovation system (NIS)

In order to implement the strategy of independent innovation and construct an innovative country, it is extremely necessary to learn from some innovative countries' experiences in building their own IS. What is an innovative country? On this issue, the international academic circles have a certain consensus that a country that takes technological innovation as its basic strategy and enhances its ability for technological innovation, and gains increasingly powerful competitive advantages, can be called an innovative country, which must have at least four characteristics as follows. The first is high innovation input. The proportion of R&D investment in the GDP accounts for more than 2% in general. Second, the rate of contribution of technological progress to economic growth is more than 70%. The third is a strong ability for independent innovation with under 30% dependence on foreign technologies. High innovation output is the fourth characteristic. At present, there are about 20 recognized innovative countries, such as the US, Japan, Finland, the Republic of Korea, etc., which have the majority of the worldwide tripartite invention patents (patents authorized by the US, Europe, and Japan). Among these countries, the methods that the US, Japan, and Finland use to create their innovation systems are especially worth learning.

(1) The construction of the US NIS

The US has a political and economic system that adapts to the market economy and a cultural atmosphere which encourages innovation. Moreover, it creates a national innovation system that is full of vitality. Its NIS has three traits: First, enterprises are clarified to be the innovation subject. At the same time, the US continues to improve its innovation-related legal system and build an environment conducive to innovation. Second, the US government only intervenes in the fields of military security and public science and technology R&D where there is "market failure," and strengthens the contact and cooperation among the government, enterprises, universities, and scientific research institutes with a large-scale research project. Third, the US attaches great importance to the cultivation and attraction of innovative talents. The US has the world's most developed modern higher education system and it continuously adjusts its immigration policy to attract the world's best intelligence resources, who provides powerful intellectual support for the S&T development.

Before World War II, the R&D in the US was mainly supported by private and corporate investment, while governmental R&D funds were very limited. The governmental policies, however, had played a great role in encouraging and supporting enterprises' innovation. First, the legislation encouraged invention and innovation. Second, European advanced technologies were introduced, implemented, and adopted in terms of both technology and talents. Meanwhile, a large number of personnel were dispatched to study in Europe, and finally, the US realized the fundamental transformation from technology introduction to independent innovation. After World War II and during the Cold War, the US government's investment in R&D increased rapidly and the funded researches realized a series of breakthroughs. Because the R&D was related closely with military hegemony, the development of S&T was greatly successful, while the gap between the development of S&T and the economic society became a problem. Since the 1980s, the US government has implemented a comprehensive innovation policy with coordination between the development of S&T and the economy, while emphasizing the enterprises' status of innovation subject and focusing on improving the enterprises' innovation ability; this policy has strengthened the cooperation among the government, enterprises, and universities.

(2) The construction of the German NIS

The German economy is export-oriented with the share of exports taking up more than 1/3 of the GDP. More than 85% of the exports come from high technology industries, including the vehicle manufacturing industry, the electronics industry, the machinery manufacturing industry, and the chemical industry. German products have become synonymous with high-quality products. Its NIS has the following characteristics.

First, the enterprise is the innovation subject and large enterprises bear most of the R&D. German R&D investment in recent years accounts for 2.54% of the GDP, which is among the best among European countries. Government investment accounts for about 35%, and enterprise investment accounts for about 65%. About 90% of the enterprise R&D investment comes from four major fields—vehicle manufacturing industry, the electronics industry, the chemical industry, and the machinery manufacturing industry. The degree of concentration in enterprise technological innovation is very high, for example, Siemens, Bosch, Infineon, BASF, and other large well-known enterprises. These enterprises are among the top 12 enterprises with the most international patent applications in the world. About 30% of German R&D is concentrated in 7 large companies, such as Siemens.

Second, the government plays an important role in encouraging enterprises to become innovation subjects. The government has proposed some policies, such as the "enterprise technological innovation risk-sharing scheme" and "the program promoting research cooperation of SMEs," which promote enterprise innovation activities.

Third, special attention is paid to the integration of the chain of industry-university-research and the strengthening of the cooperation of the government, enterprises, universities, and scientific research institutes. International Max Planck Research School and Fraunhofer Gesellschaft are unique leaders in the German NIS and have been powerful in strengthening the industry-university-research cooperation. The Max Planck is known as "the cradle of Nobel Prize." It has the departments of biology, medicine, chemistry, physics and social sciences and the humanities, and 80 research institutes, and is mainly engaged in basic research and extensive application fields through cooperation with universities. The Fraunhofer engages in applied research and undertakes research on user service contracts with industry and the government as a bridge of cooperation between universities and enterprises. At the same time, the research institutes affiliated with Helmholtz Gemeinschaft have the world's most advanced experimental equipment for scientific researchers.

Fourth, higher education and vocational training are carried out simultaneously to cultivate innovative talents effectively. There is 12-year compulsory education in Germany. Besides research-oriented universities, which pay attention to systematic theoretical knowledge, teaching and scientific research, there are also technical universities, which focus on application, that offer majors in engineering and natural sciences, and they specialize in training engineers. At the same time, vocational training is also a fundamental part of training technical talents in Germany. Students in vocational schools not only learn professional theories, but also acquire professional skills and receive related professional knowledge training. The combination of theory and technology encourages more young people to become qualified workers needed by industrial and commercial enterprises.

(3) The construction of the Finnish NIS

Finland depends on resources heavily and builds an economic system that relies on innovation and the exportation of high technology products in the three main industries of electronic information, forestry, and metal machinery through leapfrog development. The construction of Finnish NIS has at least four characteristics. The first is to pay attention to innovation at the industrial level. The institutional framework has been established to support the development of new industries. For example, in order to accelerate the development of the telecommunications industry, a series of laws and regulations related to the telecommunications industry have been formulated one after another. According to the needs of industrial development, concrete plans concerning S&T have been implemented, such as the microelectronics technology development plan. The second is to develop effectively the vocational education aiming at training highly skilled talents when emphasizing academic education. The third is the increasing investment in education and R&D. At the same time, the industry-university-research combination is promoted through technology plans and academies of science. The fourth is the rational role played by the government. Under the guidance of the concept of the construction of an NIS, the government has put forth systematic innovation policies, which promotes the improvement of innovation efficiency in the whole society.

6.2 The establishment of an NIS oriented towards independent innovation

When the comprehensive construction of an affluent society entered a crucial stage in China, according to the national conditions and the developmental needs, the Central Committee of the Party and the State Council put forward this policy—"to make S&T progress and innovation as the primary driving force for economic and social development; to make enhancing the capability of independent innovation as the key link for the adjustment of the economic structure, for the transformation of the developmental mode, and for the improvement of national competitiveness; to make building an innovative country as a significant strategy for facing the future." The authors in this book think that "system outweighs technology," so the key element in this process is the establishment of an NIS oriented towards independent innovation.

6.2.1 The historical evolution of the Chinese NIS

Since the Central Committee of the Party and the State Council established the policy for the implementation of an independent innovation strategy, the construction of an NIS oriented towards independent innovation has become the direction to which the government, and the industrial and S&T circles have channeled their efforts. In the big international environment of global economic integration, the establishment of an NIS oriented towards independent innovation suitable to the national conditions is of significance for promoting Chinese industrial development and the enhancement of its international competitiveness. Reviewing the historical and evolutionary process of the Chinese NIS and learning from its experience and lessons, we can build an NIS oriented towards independent innovation soundly and rapidly.

(1) The period of the late Qing Dynasty (1860–1911)—The incubation of the Chinese NIS

After the Opium War, with the invasion of foreign capitalism, China became a semi-feudal and semi-colonial country and modern industry appeared. Commodity output was the main tool of the foreign economic invasion, and some foreign countries forced China to sign unequal treaties and then obtained predatory trading privileges in China and attempted to gradually expand the Chinese market for their commodity. At the same time, the extravagance of the corrupt Qing government aroused people's anger, which resulted in the outbreak of the Taiping Heavenly Kingdom Movement in 1851. The specific historical conditions gave birth to the Westernization Movement. From the 1860s to the 1890s, the Westernization Group started the Westernization Movement with the slogan—"to learn from the advanced technology in the West in order to become strong." China introduced foreign technology on a large scale for the first time. The Westernization Group advocated using capitalist industry and commerce to safeguard the feudal rule of the Qing government. Historically, that was the incubation of the Chinese NIS.

1) The development of the subjects of innovation and the influence of relative systems

First, the development of industrial enterprises and relative systems. Most modern Chinese industries received investments in foreign capital in the early stages. It is estimated that 103 foreign enterprises were founded in China before 1895, respectively from Britain, Germany, Russia, France, Japan, and other countries. The majority of these enterprises were very small, and began to use power machine in mining, shipbuilding, tea processing, silk reeling, and import & export processing. At the same time, the Westernization Group took advantage of advanced technology and equipment to set up a series of new enterprises in the military industry, the civilian industry and mining as well as transportation. Between 1861 and 1895, the government funded 19 military industrial enterprises, 29 civilian industrial enterprises, and a group of transportation enterprises, and built 364 kilometers of railway. The machines in the military factories were all ordered from Britain, France, the US, Germany, and other countries. The positions of technicians and foremen in large arsenals and shipyards were occupied by foreigners. Since 1875, the Qing government began to operate in the mining, iron smelting, and textile industries and had built more than 10 enterprises by 1895. From the 1870s, some Chinese merchants, landlords, and bureaucrats began to invest in modern industry, and by 1895 more than 100 enterprises had been established. Generally, these enterprises were on a smaller scale, but they laid the foundation for the modern industries funded by national capital. Since 1895, the number of Chinese capital manufacturing enterprises had significantly increased. During the period 1895–1913, at least 549 merchant-operated and semi-official manufacturing and mining enterprises that used machine power were established.

Second, the development of education and the influence

of relative systems. In the 1860s, the Westernization Group began to take education as the major project, which ranks only second to the military industry. They advocated the policy—"Chinese Learning as the Fundamental Structure and Western Learning for Practical Use." The reform of the educational system dominated by the government mainly included setting up new-style schools, sending students to study abroad, translating Western books, and taking the initial attempt at reforming the feudal system of education.

First of all, the reform involved setting up a series of new-style schools. From 1862 to the end of the 19th century, the Westernization Group successively established more than 30 schools in all including foreign language schools, military schools, technical schools, etc. These schools laid a preliminary foundation for the formation and development of higher education in modern China. Another measure of the Qing government's educational reform was to send students to study abroad. In 1872, the Qing government for the first time designated Rong Hong to lead a group of 30 children to study in the US. By October 1875, the Qing government had sent a total of 4 groups of children to the US. From 1877 to 1897, 84 students from Fuzhou Shipping School were sent to Europe to study. Most students in the US were studying natural science or applied science, including electricity, medicine, mining, ironware, civil engineering, and railway construction. The students in Europe were mainly studying Western shipbuilding and driving technology, so as to strengthen the coastal defense. The educational reform of the Westernization Movement, to a certain extent, promoted the spread of Western learning in China and cultivated a batch of science and technology, business, military and diplomatic talents.

In 1898, the Reformers launched the Reform Movement aiming at making the country stronger through reform. The Reformers thought that China should learn not only Western science and technology, but also Western political system; they advocated abolishing the stereotyped essay, reforming the imperial examinations, setting up schools all over the country, learning Western science, rewarding inventions and books writing, etc. They thought that the development of education was of vital importance to the prosperity of the country. The Reformers set up the Imperial University of Peking in the same year, which opened up Chinese higher education. It established a complete set of subjects to be taught, which contained economics, law and politics, liberal arts, science, engineering, business, and agriculture. At the same time, running schools became very popular. By 1904, 4,222 schools had been founded with 92,169 students. In 1909, the number of students reached 1,560,270, and the number of schools reached 52,348.

2) Interaction and the interactive relationship among the subjects of innovation

The Westernization Movement during this period not only gave birth to modern Chinese industry, but also promoted the reform and development of Chinese education and cultivated a large number of talents, which laid the intellectual foundation for the construction of an NIS. However, during this period, the development of Chinese technology relied on the importation of technology; the scale of enterprises funded by national capital was small; they were mainly engaged in light industry and small-scale mining; machinery and equipment were totally dependent on importations from abroad and principal technicians were foreigners. Western-style education obtained a certain number of achievements during this period, and cultivated a batch of talents with advanced technology. And the majority of these people were sent to the official enterprises, which stimulated the development of capitalist industry and commerce. However, there was no interaction regarding innovation between enterprises and educational institutions.

The relationship among the subjects of innovation during the period of the late Qing (1860–1911) is shown in Figure 6-1.

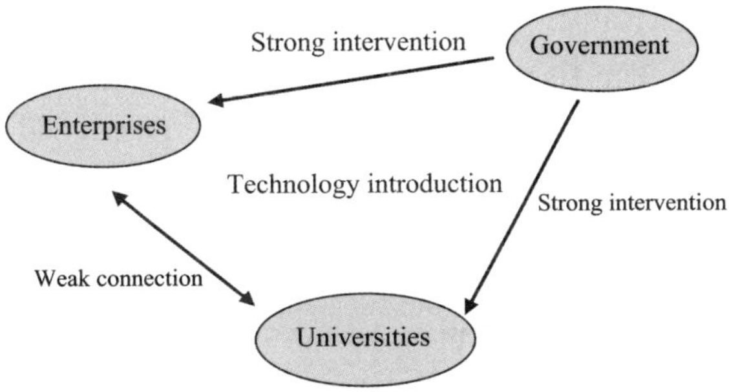

Figure 6-1 The relationship among the subjects of innovation during the period of the late Qing Dynasty (1860–1911)

(2) The period of the Republican China (1912–1949)—The germination of the Chinese NIS

The Republican China was a period of a huge economic, social, and cultural revolution. However, from 1912 to 1949, China experienced slow economic growth, and the production was mainly from agriculture. The proportion of the urban and rural population had not changed significantly. There were few financial institutions. A unified national market had not formed.

1) Enterprise development and the influence of relative systems

At the end of the 19th century and the beginning of the 20th century, capitalism started to develop in China. National industries that developed during this period were mainly light industries, such as the cotton textile industry, the flour industry, the silk reeling industry, and the papermaking, cigarette, oil, and sugar industries, etc. In heavy industries, because coal and iron production was mostly manipulated by imperialist countries, the growth of national iron and coal production was very limited. From 1927 to 1937, the national government set up the National Construction Committee, the National Economic Committee, and the National Resources Committee. This action played an important role in economic and social development, such as the implementation of the five-year plan of heavy industrial construction, the realization of tariff autonomy and monetary reform, the implementation of the nationalization of the railway, the construction of the Zhe-Gan, Yue-Han, Long-Hai, and Huainan railway lines (9,000 km in total). At the same time, the national government had established a financial system, implemented a financial monopoly, and controlled industry, mining, and transportation. In the industrial enterprises of this period, Chinese capital enterprises accounted for about 70% of industrial output and foreign capital accounted for 30% (1933). The national capital enterprises were mainly light industry enterprises, 50% of which were engaged in textile and food production. From 1933 to 1937, national industry developed greatly. By 1937, there had been a total of 2,435 Chinese-owned factories with 493,257 workers, which were concentrated mainly in Shanghai and other coastal areas. In 1943, there were a total of 3,758 factories, 50% of which produced military supplies and capital goods. However, during this period, the first disadvantage was that the average factories were very small in size, and were mainly concentrated in the coastal cities, and the average number of workers in the factories was around 200; the second disadvantage was that the factories had backward technology, the machinery was purchased from abroad, and the factories rarely used mechanical power and complicated machinery; the third disadvantage was the general lack of technical and managerial personnel. For example, in 1931, in 82 cotton mill enterprises, only 500 out of the 4,000 technicians had received formal training. Nevertheless, in general, before 1949, Chinese enterprises lacked any kind of basic ability for innovation (Fei Zhengqing, 2006).

2) Educational development and the influence of relative systems

During the period of the Republican China, Chinese higher education developed to a certain degree. In 1922, the Ministry of Education of the Beiyang government published the Project for the Reform of the School System, announcing the implementation of a new educational system and the reform of the higher educational system. In 1924, the 20 Regulations of the National University were published, which proposed that the "national universities should aim at teaching advanced academics, cultivating talents to meet the requirements of the country." (Qu Tupei, 2006: 268-276) The "French-style educational system" was implemented by the national government in 1927. The Graduate School established by the central government is the country's highest academic and educational institution. The university & college and the university district system was implemented, but this system died young in less than two years. After that, the national government announced the University Organization Law, the University and College Regulations, the Technical College Organization Law in 1929, and the Technical College Regulations in 1931. By 1937, before the outbreak of the Anti-Japanese War, 108 universities and colleges had been established in the Kuomintang-controlled areas, including 42 universities (13 national universities, 9 provincial and municipal universities, and 20 private universities), 36 independent institutes (5 national, 9 provincial and municipal, and 22 private), and 30 technical colleges (8 national, 11 provincial and municipal, and 11 private). The universities had eight kinds of schools: Literature, science, law, engineering, agriculture, business, medicine, and education. Most schools had libraries and laboratories, and professors were engaged in academic research and teaching simultaneously. The Central Research Institute and the Academy of Peking were set up during this period, and there were 12 universities which had research departments. The number of undergraduate students and graduate students of universities and research institutes reached more than 40,000 (Chen Dongyuan, 1948: 525, 1406). Overall, the main task of universities and colleges was the popularization of higher education, personnel training, but cultivating research talents and academic research was still not its main function.

3) The development of scientific research institutes and the influence of relative systems

The Chinese scientific research system began early in the 20th century. Around 1915, the Science Society of China was established. Subsequently, in 1916, the first public research institute, the Central

Geological Survey, was established. In 1922, the Biological Research Institute was established in Nanjing. In 1928, Jingsheng Biological Survey was established in Beijing. The Central Research Institute was established in the same year. In 1929, the Academy of Peking was established. However, during this period, affected by war and social unrest, Chinese science and technology developed very slowly. The number of scientific and technical researchers around the country was no more than 50,000 with fewer than 500 people who specialized in scientific research in 1949. There were only more than 30 scientific research institutes, being mainly concentrated in the fields of geological science and biological taxonomy.

4) The institutional construction of the Chinese NIS

The institutional construction of the Chinese NIS in this period can be briefly summarized as in Table 6-1.

5) Interaction and the interactive relationship among the subjects of innovation

Chinese industry was dominated by the handicraft industry; the scale of enterprises was generally small; equipment was imported; enterprises' internal research institutes had not yet emerged. The main purpose of universities was to popularize higher education and cultivate talents; the universities were rarely engaged in R&D, and few universities had begun to establish research institutes. In addition, the establishment of independent research institutes, such as the "Central Research Institute," started to systematize research activities. The enterprises, universities, and research institutes were independent at that time, rarely cooperative (shown in Figure 6-2). Nevertheless, the development of local enterprises and universities as well as the emergence of scientific research institutes laid certain foundation for the later construction of the Chinese NIS.

Table 6-1 The institutional construction of the NIS during the period of the Republican China

Time	Events	Main achievements
	Scientific Research	
1919	The establishment of the Science Society of China	The earliest S&T organization
1916 – 1928	The earliest establishment of research institutes—the Central Geological Survey (1916), the Biological Research Institute (1922), and Jingsheng Biological Survey (1928)	The earliest independent scientific research institute
1928	The establishment of the Central Research Institute	The highest level of academic research institution in the period of the Republican China, under supervision of the office of the President
	Higher Education	
1922	The Ministry of Education officially announced the Reform of the School System	Reform of the educational system
1924	The Ministry of Education announced the National University Regulations	Re-enacting the relevant system of higher education
1929	The Ministry of Education officially announced the University Organization Law, the University and College Regulations and the Technical College Organization Law	The first issued educational regulations
	Economy	
1928	The government set up the National Construction Committee	The highest administrative organ in industrial construction
1931	The government set up the National Economic Committee	The highest administrative organ in the construction of the economy
1935	The government set up the National Resources Committee	To investigate and develop national defense resources and make a massive investment plan, the five-year plan of heavy industry construction was formulated in 1935

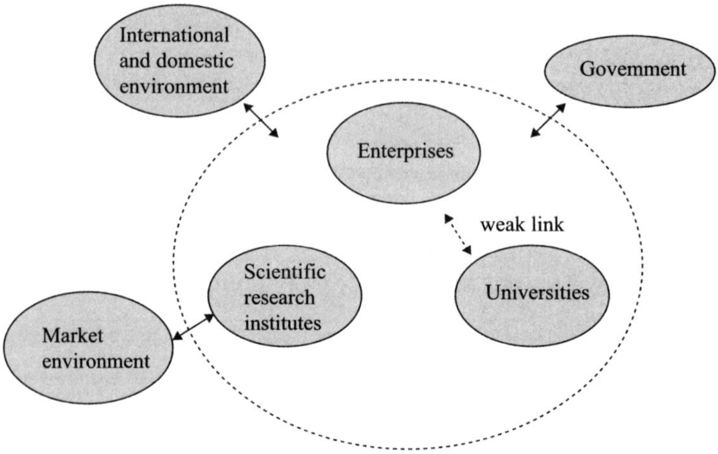

Figure 6-2 The relationship among the subjects of innovation during the period of the Republican China

(3) From the establishment of People's Republic of China to the implementation of reform and opening up (1949–1978)—The construction of the Chinese NIS

When People's Republic of China was founded, there was a sharp confrontation between the East and the West—the alliance of the Soviet Union and China was against the West. During the period from 1949 to 1952, the economy of China began to recover: The government unified the finance and the economy, completed the adjustment of industry and commerce and the recovery and development of agriculture. In 1953, China began to implement the first five-year plan for realizing the transition to socialism. The state carried out the socialist transformation of agriculture, the handicraft industry, as well as industry and commerce. After the accomplishment of the first five-year plan, the preliminary basis of the socialist economic system and industrialization was established. In 1958, the second five-year plan was carried out. Industry developed rapidly with an average annual production growth of 18%. The period from 1958 to 1976 was turbulent for the socialist economy, namely the "Great Leap Forward," the People's Community Movement, and the "Cultural Revolution" from 1966 to 1976; the national economy was totally out of order. Looking at the whole situation, from the foundation of People's Republic of China to 1977, China had established many emerging industrial departments, such as electronics, aviation, petrochemical, automobile, and atomic energy. Considering the industrial structure, China had developed from the domination of the first industry to the 50-50 split of the first and the second industry. China developed gradually from an agricultural country to an industrial society, and a complete industrial system was basically established. At the same time, a relatively complete S&T system was established. Universities and scientific research had developed quite satisfactorily.

1) The development of the subjects of innovation and relative systems

First, industry and its development. At the beginning of the foundation of People's Republic of China, its economy was very backward, with the industrial output value accounting for only 17% of the total output value of industry and agriculture and the net value accounting for only about 10% of the total. At that time, enterprises were mainly made up of state-owned enterprises and private enterprises with the latter in the dominant position. Capitalist industrial enterprises reached more than 123,000 with more than 1,640,000 staff, which accounted for 54.6% of the total. The industrial value of private enterprises accounted for 48.7% of the total. In 1954, the state announced the Provisional Regulations for Public-Private Partnership of Industrial Enterprises, which officially recognized the status of a public-private partnership economy; and the enterprises with public-private partnerships developed rapidly from 193 in 1949 to 3,193 in 1955, whose production accounted for 16.1% of the gross industrial value.

After the accomplishment of the reform of capitalist industry and commerce, the state-owned economy gradually occupied the dominant position. Enterprises took the national mandatory production plan as their goal. The main function of enterprises was carrying out production according to the arrangement of their superiors. Thus, enterprises lacked the autonomy and initiative to promote technological innovation.

There were two ways for enterprises to acquire production resources to improve production. The first one was to transfer the technical achievements from the R&D agencies belonging to the government through government department in charge; the second one was to import equipment or production lines from abroad with the approval of the government department in charge. Under this system arrangement, enterprises could not get innovation profit and did not need to bear the loss of innovation failure. Although internal technical reconstruction of enterprises during this period was quite intensive, it was the result of the planned arrangement of the government department in charge. Military enterprises obtained the most rapid development in this period. Since the foundation of new China, it had been facing the threat of American imperialism. In order to strengthen national defense as soon as possible, China learned from the Soviet Union, while introducing their technology, so that the development of military enterprises and related enterprises got promoted. Then, following the test explosion of the first atomic bomb in October 1964, the test explosion of the first hydrogen bomb successfully took place in June 1967. In 1969, the first underground nuclear test was successfully conducted. In 1970, the first man-made earth satellite was successfully launched. In 1975, the man-made earth satellite returned to the ground successfully, and China became the third country able to successfully recover a satellite after the United States and the Soviet Union. In addition, during this period, China also developed a large number of major new products and complete sets of equipment, such as 30,000-ton forging hydraulic press, etc.

Second, the development of universities and relative systems. After the foundation of People's Republic of China, especially during the first five-year plan period, Chinese cultural and educational undertakings developed to a great extent. In 1953, there were 184 universities and colleges in total, including 14 comprehensive, 38 industrial, 33 normal, 29 agricultural and forestry, 29 medical, 6 financial, 4 politics and law, 8 language, 15 art, 4 sports,

3 ethnic, and 1 of other kind of institute. In 1957, there were 441,000 students in colleges and universities, which had increased by 130.9% from what it was in 1952 (191,000) and by 184.5% from the number in 1949 (155,000). The "Great Leap Forward" began in 1958; the number of universities and colleges had increased during these three years to 1,289 in 1960. Then, the state made adjustment in the universities and colleges. The number of universities and colleges dropped to 407 in 1963, the number of students diminished from 961,623 in 1960 to 750,118 in 1963. Higher education was on the road to healthy development again, and the quality of the teaching had obviously improved. However, the "Cultural Revolution" from 1966 to 1976 seriously damaged the educational sector, and higher education was the first to be interrupted.

Looking at the situation as a whole, during that period, the development of Chinese education focused on cultivating talents. The "Common Program of Chinese People's Political Consultative Conference" formulated in 1949 ruled that: "The work on culture and education of the people's Government is supposed to improve people's cultural level, cultivate talents for national construction, eliminate the feudal, comprador, fascist ideology, and develop the ideology of serving the people." Another characteristic of universities at that time was the emphasis on professional development in the industrial field and a neglect of law and finance. Later, the Ministry of Education carried out a significant reform of higher education, drew the lessons from the experiences of the Soviet Union to carry on adjusting the structure of universities and colleges. The policy of adjustment was "with the purpose of cultivating cadres and teachers for industrial construction, developing specialized colleges and technical colleges, and rectifying and strengthening comprehensive universities." In 1953, the structural adjustment of the universities had been almost completed.

Third, the development of scientific research institutes and relative systems. At the beginning of the foundation of

People's Republic of China, there were only a few more than 700 S&T talents and more than 30 specialized research institutes. Geological and biological classification research were equipped with some of the basics. Modern technology was almost void. After the foundation of People's Republic of China, the Party and the government attached great importance to the development of science and technology. In 1949, the Chinese Academy of Sciences was established, and in the next few years, S&T coordination and research institutes, such as China Association for Scientific and Technical Workers, China Meteorological Administration, the Ministry of Geology, were established. In 1956, the State Council set up the National Science Planning Committee, formulated the Draft of Science and Technology Development, and put forward 12 key projects with 57 important science and technology tasks. On this basis, the State Science and Technology Commission was set up in 1958, followed by the establishment of the Science and Technology Work Committee for National Defense. Thus, the science research system was initially formed made up of the Chinese Academy of Sciences, universities, national defense research institutes, institutes of industrial departments, and local academy of sciences under the guidance and coordination of the State Science and Technology Commission, the Ministry of Education, the Science and Technology Work Committee for National Defense, and other organizations. By 1957, the number of scientific research institutes had reached more than 500; the number of researchers was more than 20,000, two times more than that in 1952; from then until 1966, scientific research institutes grew rapidly.

The main characteristics of the development of the Chinese scientific research system during that period can be described as follows: Research institutes carried out scientific research under governmental orders, and S&T funds were allocated by the government; research institutes were completely independent of enterprises; the government focused on the S&T development of national defense to ensure the smooth running of major S&T plans with military objectives; but the

R&D of scientific institutes was disconnected with the production practice of enterprises, leading to the failure of the transformation of S&T achievements into productivity.

Fourth, the institutional construction of the Chinese NIS. From the foundation of People's Republic of China until the implementation of reform and opening up, under the leadership of the central government, the work related to the institutional construction of the Chinese NIS can be summarized as in Table 6-2.

Table 6-2 The institutional construction of the Chinese NIS from the foundation of People's Republic of China to the implementation of reform and opening up

Time	Events	Main functions
	Science and Education	
1949	The establishment of the Chinese Academy of Sciences	Scientific research in every field of natural and social science
1949	The foundation of China Association for Scientific and Technical Workers	Unifying workers in natural science and technology, and consultation on major S&T issues
1956	The establishment of the National Science Planning Committee	The permanent organization responsible for national S&T work
1958	The foundation of the State Science and Technology Commission	Responsible for the coordination of the national S&T activities, and technology planning and development strategy
1955	The State Council approved the Provisional Regulations on the Bonus of the Chinese Academy of Sciences	The establishment of the reward system for science and technology
1955	The establishment of the system of postgraduate training—Provisional Regulations of Postgraduate Training of the Chinese Academy of Sciences	Focusing on the development of higher education and improving the quality of talent training
	Industry	
1953	The formulation of the first five-year plan	Restoring industrial construction and focusing on developing heavy industry

(Continued)

Time	Events	Main functions
1954	The state announced the Provisional Regulations on the Public-Private Partnership of Industrial Enterprises	The formal establishment of public-private economic status

2) The interaction and relationship among the subjects of innovation

From the foundation of People's Republic of China to the implementation of reform and opening up, China implemented a highly centralized planned economic system. In this context, the relationship among the various subjects of innovation was very distant, and enterprises and research institutes were completely independent. Enterprises only produced goods, but rarely set up R&D institutes, let alone engaged in independent R&D activities. Research institutes took R&D as their main task which were arranged by the government, and R&D achievements were transferred to specific enterprises according to governmental plans. Universities took the cultivation of talents as their main task, and scientific research served teaching, which was rarely transferred to the enterprises. All these resulted in a serious gap between R&D and enterprise production practice.

The framework of the NIS from the foundation of People's Republic of China to the implementation of reform and opening up is shown in Figure 6-3.

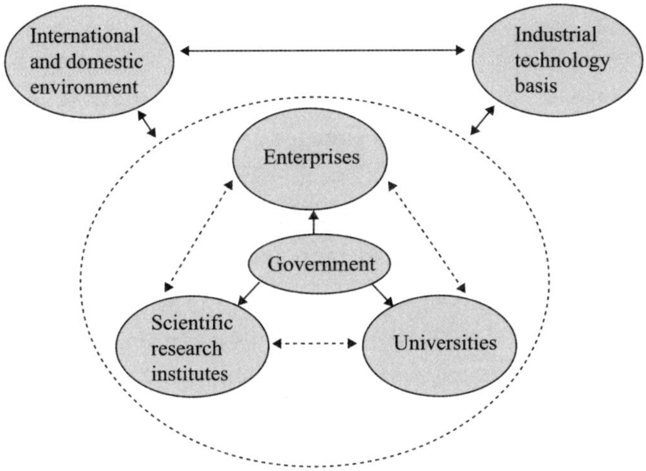

Figure 6-3 The framework of the NIS from the foundation of People's Republic of China to the implementation of reform and opening up

(4) After the implementation of reform and opening up (1978–2005)—The gradual formation of the Chinese NIS

After the Third Plenary Session of the 11th Central Committee of the Chinese Communist Party, China began to implement the reform of the economic system; the planned economy system began to change towards the market economy system. During that period, the Chinese industrial foundation was increasingly strengthened; major changes took place in the structure of industry, the proportion of secondary and tertiary industries increased clearly, the level of science and technology improved, and the national economy grew steadily and sustainably.

1) The development of the subjects of innovation and the influence of relative systems

First, enterprise development and the influence of relative systems. Since the implementation of reform and opening up, great changes have taken place in the ownership structure of Chinese enterprises, which has evolved gradually into the coexisting situation of state-owned enterprises, collective enterprises, private enterprises, joint ventures, foreign-owned enterprises, etc. The enterprise scale and quantity have grown substantially. The number of all sorts of LMEs reached 14,935 in 1991, and it reached 40,314 in 2008. It increased by 2.699 times in fifteen years. The strength of the whole enterprise R&D greatly improved; the number of R&D personnel grew substantially; the proportion of the personnel engaged in S&T activities in the total personnel increased from 0.382% in 1992 to 0.637% in 2008; the proportion of R&D personnel increased from 0.113% in 1992 to 0.252% in 2008. S&T expenditure also increased significantly from 84.69 billion yuan in 1995 to 851.06 billion yuan in 2008, namely a tenfold increase. Enterprises gradually became the subjects of innovation in market economy, and many LMEs set up their own research institutes.

Another characteristic of enterprise development is that the enterprises' ability to carry out independent R&D was not strong with not much investment, and the expansion of production capacity mainly depended on importing key equipment from abroad. For example, despite the fact that the proportion of the R&D expenditures of LMEs in the main business income increased from 0.49% in 1991 to 0.84% in 2008, compared with developed countries, there was still a big gap. At the same time, enterprise R&D investment was far lower than investment in the import of technology. For example, in 1995, the LMEs spent 38.636 billion yuan on purchasing and importing technology, but the R&D expenditure was only 14.17 billion yuan, far less than the expenditure on the purchase of technology. In the two types of expenditure, namely the import of technology and the implementation and adoption of technology, the investment in latter was obviously lower than the former. For example, in 1995, the expenditure of the latter was 1.31 billion yuan, accounting for only 3.6% of the expenditure of the former, far lower than the proportion in developed countries (for example, in Japan, the long-term proportion remained at about 7%).

In recent years, it is increasingly clear that enterprises are the innovation subject in market economy. First, enterprise R&D expenditure increased significantly and reached 321.16 billion yuan in 2008. Second, the relationship between the expenditure on the importation of key equipment and that on the implementation and adoption of technology tended to be harmonious. Both of them reflect that with relevant governmental policies and enterprises' own efforts, the enterprises' status as the subject of innovation, especially that of independent innovation, got promoted in the market economy.

Second, the development of universities and the influence of relative systems. After the implementation of reform and opening up, Chinese higher education developed in an unprecedented manner. In 2000, there were 1,041 regular universities and colleges with 949,000 undergraduate students and 58,800 graduate students. In 2005, the number of universities and colleges increased to 1,792, and there were 3,067,000 undergraduates and 189,700 graduate students.

The main characteristics of the development of universities and colleges in that period are that the universities and colleges not only cultivate a large number of personnel for the country, but also play an important role in the NIS to bear a lot of research work. First, the number of the research institutes created by universities and colleges increased steadily. By 1995, the universities and colleges had set up 3,421 R&D institutes, and the number had reached 5,159 by 2008. Second, the number of professional R&D personnel in universities and colleges increased significantly from 324,000 in 1995 to 579,700 in 2008. Third, the S&T expenditure increased significantly from 4.44 billion yuan in 1995 to 59.31 billion yuan in 2008, increasing by 13.4 times within ten years. Fourth, the distribution of R&D investments in universities and colleges was more reasonable in basic research, pilot development, and applied research. The proportion of basic research funds increased year after year, from 15.4% in 1995 to 29.4% in 2008; the proportion of pilot development funds decreased year after year from 29.6% in 1995 to 17% in 2008; the proportion of applied research remained stable within the range from 49% to 55%. However, compared with developed countries, the basic research work in universities and colleges was still weak, and this may be a reason for fewer original innovation achievements in China. Fifth, the universities and colleges paid more attention to the commercialization of S&T achievements. Especially, the cooperation between universities and enterprises was increasingly frequent. The technology achievement transfer and patent sale from universities and colleges to enterprises can provide some proof of it. During that period, the technology transfer contracts signed by universities and colleges and enterprises increased from 4,030 in 1992 to 6,878 in 2006; patent sales increased from 486 in 1992 to 701 in 2006.

Since the implementation of reform and the opening up, these changes in universities and colleges can be attributed to two factors. The first is that the government makes the three functions of universities and colleges substantially clear (i.e. personnel training, scientific research, and social service); the second is that since the first National Innovation Conference, the government has launched relevant policies to promote scientific R&D and the transformation of S&T achievements.

Third, the development of scientific research institutes and the influence of relative systems. Since the implementation of reform and opening up, Chinese professional scientific research institutes had changed a lot. First, fiscal allocation for scientific research institutes from the government increased significantly. For example, the number in 1995 was 17.92 billion yuan and 115.6 billion yuan in 2008. Second, the number of professional research institutes decreased gradually. In 1995, there were 5,841 professional R&D institutes, the number of which diminished to 3,727 in 2008. Part of the professional research institutes turned into high-tech enterprises. Third, the proportion of basis scientific research in professional scientific research institutes increased year after year. The proportion of basic research expenditures increased year after year from 7% in 1995 to 11.4% in 2008; the proportion of applied research increased slightly from 31.1%

in 1995 to 33.4% in 2008; the proportion of pilot development declined annually, which decreased from 62% in 1995 to 55.1% in 2008. This is reflected in the fact that the R&D activities in professional research institutes changed gradually from an emphasis on pilot development to basic and applied research.

These changes in professional research institutes are attributed to the following three reasons: The first one is the effective implementation of the major principles advocated by the government—"scientific research institutes should find research topics from the realistic development needs of the national economy and serve it." The second reason is the adjustment in the industrial scientific research system since the first National Innovation Conference. Especially, part of the industrial scientific institutes turned into high-tech enterprises and the number of professional research institutes decreased. Third, the Chinese Academy of Sciences is the implementation subject of the National Knowledge Innovation Project. The fiscal investment in basic research and applied research increased.

Fourth, the development of the enterprise incubator and the influence of relative systems. Since the implementation of reform and opening up, the enterprise incubator had become an important part of the Chinese NIS, playing a unique role in promoting innovation and entrepreneurship. The first Chinese incubator was established in 1987. By the end of 2004, there had been more than 500 incubators of different types and more than 40,000 incubating enterprises. The creation and development of the enterprise incubator in China mainly benefited from the Torch Plan after the implementation of reform and opening up, from the relevant policies of "promoting the development of high and new technology industries" after the First National Conference on Science and Technology, and from the effective implementation of the relevant policies of the "vigorous promotion of independent innovation" after 2006.

2) The institutional construction of the Chinese NIS

From the implementation of reform and opening up until the year 2000, the institutional construction of the Chinese NIS can be divided into two stages.

The first period is from 1978 to 1992. The transition from the planned economic system to the socialist commodity economic system was the key element during this period. Profound changes took place in the institutional construction of the Chinese NIS. Especially, the idea that "science and technology is the first productivity" was admitted and gradually implemented in practical work, which played a fundamental role in the institutional construction of the NIS in the new period. At the governmental level, first, the Regulations of Science and Technology Progress Award (1984) was issued and the award was set up. Second, the National Natural Science Foundation (1986) was established. The national funding system for basic and applied research was also established. Third, the Trademark Law (1982), the Patent Law (1984), and the Copyright Law (1990) were officially announced in succession, which led to the initial construction of an intellectual property system in China. Fourth, the Technology Contract Law (1987) was announced, which provided a legal basis for the adjustment of the relationship between civil rights and obligations that involved technological development, technological transfer, technological advice, and technological services. At the micro level, first, scientific research institutes and colleges and universities are encouraged and supported to enter the economic field directly in various forms, and a new market-oriented scientific research production and operation entity was established; second, the reform in professional scientific research institutes was intensifying. At the same time, the government actively supported the development of scientific research institutes of collective ownership, private ownership, and ownership in other forms. These policies and measures made the Chinese NIS develop substantially. Though during that period, the "decentralization" enhanced the vitality of enterprises, still mainly depending on scale expansion to strive for better development, and coupling with the short-term practice of the "contract system," they lacked the inherent motive behind innovation. In other words, enterprises were still not the real subjects of innovation.

The second period is from 1992 to 2005. In 1992, in his "Talks in the South," Deng Xiaoping clearly stated that the basic goal of the reform of the Chinese economic system was to establish a socialist market economy. From then on, China was gradually transformed from a socialist commodity economic system to a socialist market economic system. At the same time, the strategies of "revitalizing the nation through science and education" and "sustainable development" were implemented, a series of reforms were implemented concerning the institutional construction of the NIS. At the governmental management level, the first reform was the legislation regarding science and technology activities. The Science and Technology Progress Law (1993) and the Scientific and Technological Achievements Transformation Law (1996) were promulgated. Second, the top organization construction of the S&T development was strengthened. The National Leading Group of Science and Education (1998) was set up, which was in charge of coordinating and leading the work of national S&T and education. Third, the administrative departments of S&T and the economy cooperated in formulating a major S&T plan and promoting the combination of S&T with the economic project and plan at the national level. At the micro level, the first was that the construction of a modern enterprise system was sped up and the inner motive of enterprise innovation was strengthened; second, the national engineering center, the industrial technology center, and the productivity promotion center beneficial to enterprise innovation were established to provide an effective support for animating the activities of enterprise innovation.

The institutional construction of the Chinese NTS from the implementation of reform and opening up until the year 2000 is shown in Table 6-3.

Table 6-3 The institutional construction of the Chinese NIS from the implementation of reform and opening up until the year 2000

Time	Events	Main Effects
	Science and Education	
1978	The establishment of the Chinese Academy of Social Sciences	Research and advisory activities in various fields of the humanities and the social sciences
1984	The promulgation of the Regulations of Science and Technology Progress Award	Setting up a scientific and technological progress award, which is divided into the national and the provincial scientific and technological progress award with specific provisions of grade, medal, and money
1986	The establishment of the National Natural Science Foundation	To support basic and applied research
1990	The promulgation of the Copyright Law	Aiming at the protection of IPR
1993	The promulgation of Science and Technology Progress Law	The state will gradually increase S&T funds and set up an S&T award system
1996	The promulgation of the Scientific and Technological Achievements Transformation Law	
1998	The establishment of the National Leading Group of Science and Technology	Coordinating and leading national S&T and education
1999	The promulgation of the Science and Technology Awards Regulations	Setting up Highest Science and Technology Award, National Natural Science Awards, National Technology Invention Awards, National Science and Technology Progress Awards, as well as China International Science and Technology Cooperation Award
	Enterprise	
1982	The promulgation of the Trademark Law	A registered trademark is valid for 10 years from the date of the approval of the registration in order to ensure the rights and interests of enterprises and encourage their innovation
1984	The promulgation of the Patent Law	Specific measures for the protection of patent rights
1987	The promulgation of the Technology Contract Law	It involves the relationship of civil rights and obligations of technological development, technological transfer, technological advice, and technological services

3) The relationship among the subjects of innovation in the Chinese NIS

From the implementation of reform and opening up to 2005, the interaction among various types of innovation subjects in China, especially the cooperation among industry, universities, and research institutes was gradually strengthened. The industry–universities–research institutes cooperation in our country began basically from the beginning of the 1980s. In 1986, the former National Economic Commission, National Education Commission, and the Chinese Academy of Sciences jointly advocated the implementation of the "united development project of industry–universities–research institutes" to accelerate the transfer of the scientific achievements of universities to enterprises and to improve the enterprises' ability for innovation. Up until August 2003, the number of united development agencies of cooperation among industry–universities–research institutes had reached more than 6,200; and more than 130,000 enterprises had become members of the alliance of cooperation among industry–universities–research institutes. The number of universities and research institutes in the project was more than 8,000. They had developed or transferred more than 150,000 technologies. At present, the cooperation among industry–universities–research institutes in the Chinese NIS has become closer, which is embodied as follows.

First, dynamic joint research activities. The joint research activities of universities and enterprises have become increasingly active. From 1985 to 1997, the proportion of the scientific research funds of universities from enterprises had been increasing year after year, and it rose from 6.18% in 1985 to 49.24% in 1997; the proportion of governmental allocation had been gradually declining, falling from 71.45% in 1985 to 45.37% in 1997. In 1990, among all the research topics of R&D institutes, those entrusted by enterprises accounted for 14.89%; their personnel input accounted for 12.22% of all. By 1996, the proportion of these two had increased to 16.47% and 15.30%, respectively. The number of domestic scientific papers jointly published by enterprises and scientific research institutes had an average annual growth rate of 5.52% from 1992 to 1997; the number of domestic scientific papers jointly published by enterprises and universities had an average annual growth rate of 14.88% from 1992 to 1997.

Second, joint patent applications. The number of joint applications for patents by enterprises, research institutes, and universities had been increasing. From 1986 to 1993, the number of joint applications for patents by enterprises and research institutes increased with an average annual growth rate of 39.41%. The number of that by enterprises and universities increased greatly from 1986 to 1988, but were stable from 1988 to 1997. Joint patent applications among enterprises were the most active, the number of authorized patents increasing from 8 to 213 with an average annual growth rate of 34.76%. However, during that period, the inter-institutes, and universities and research institutes together made few joint applications for patents.

Third, technological transfer. Enterprises have gradually become the biggest buyers of technical achievements from universities and institutions. The technical contract amount with the enterprises as the buyers was 0.626 billion yuan in 1986 and 5.102 billion yuan in 1997, which means an average annual increase of 21% and this accounted for around 73% of the total value. From 1985 to 1997, the transfer amount of technology from universities to enterprises increased from 0.13 billion yuan to 0.65 billion yuan with an average annual growth rate of 9%.

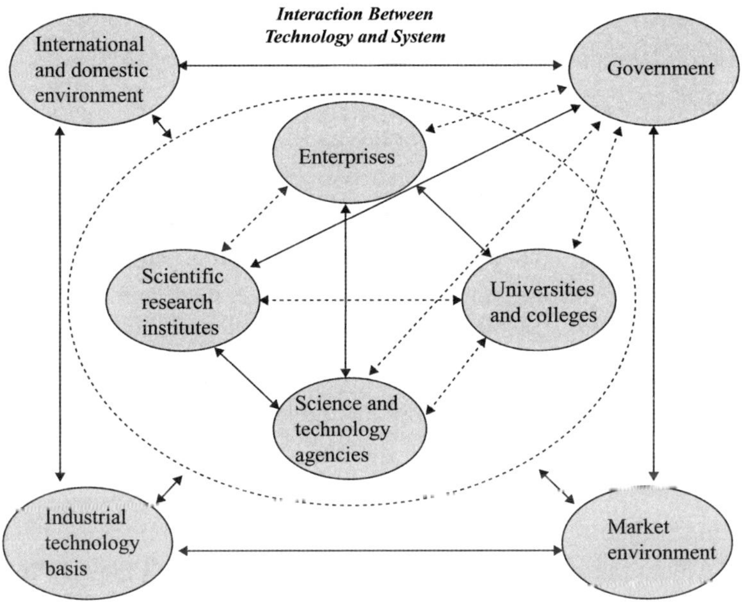

Figure 6-4 The basic framework of the Chinese NIS from the implementation of reform and opening up to 2005

The basic framework of the Chinese NIS from the implementation of reform and opening up to 2005 is shown in Figure 6-4.

6.2.2 The construction of the Chinese NIS oriented towards independent innovation

(1) The great reform of the Chinese NIS after 2005

From January 9 to 11, 2006, China held the first National Science and Technology Conference in the 21st century, and clearly put forward the major strategy of "taking the road towards independent innovation with Chinese characteristics and building an innovative country." Moreover, the implementation of the Outline of the National Medium and Long-term Science and Technology Development (2006–2020) pointed out the direction for the development of science and technology. and the construction of the Chinese NIS. "System outweighs technology." It means that by taking the road towards independent innovation and building an innovation-oriented country, we need to construct an adaptable NIS, that is, an NIS oriented towards independent innovation.

Historically, the Chinese NIS since the implementation of reform and opening up has been typically oriented towards imitative innovation. First, considering the behavior preferences of the subjects of innovation, the higher education prefers to imitate the West and is keen on following and imitating education in developed countries; scientific research institutes prefer to imitate foreign R&D and are eager to fill the gaps; enterprises lack the R&D ability, and coupled with the preference for immediate interests, it is very natural that they are keen on importing or imitating a large quantity of the innovation achievements of leading foreign or domestic enterprises. Second, as for innovation input and its mechanism, R&D investment from the entire Chinese society is not sufficient. Taking the proportion of a country's R&D input in the total global R&D in 2002 as an example, it was 41.7% in the US, 22.3% in Japan, 2% in the Republic of Korea, and only 1.7% in China. There was a low proportion in basic research investment in China and it was on a small scale. In 2002, the proportion was 16.6% in the US, 14.4% in Japan, 16.1% in Russia, 17.6% in India, 14% in the Republic of Korea, and 5.7% in China (before 2002, it was only 2.3%). Independent innovation is essentially technological innovation based on knowledge innovation. When basic research investment has a low proportion and is on a small scale, it is difficult to carry out knowledge innovation. The lack of knowledge innovation leads technological innovation to become "water without a source," and the whole society has to naturally only rely on imitative innovation. Third, considering the incentive mechanism of management culture, the "No-Fault Management" is popular in various organizations (enterprises, universities, institutes, and governmental agencies). People can have no achievements, but they cannot make mistakes. Under such management concept, the members of these organizations were naturally reluctant to take risks in independent innovation. Coupled with the large market space for "imitation rent-seeking," enterprises were naturally induced not to carry out independent innovation, but to prefer imitative innovation instead. Fourth, regarding the orientation of governmental policy, under the condition that the "market for technology" policy had been implemented for a long time after the implementation of reform and opening up, the market had not gotten any core technology in return, which also made enterprises rely on imitating foreign technology. In particular, under the great employment pressure and the responsibility and pressure to improve people's living standard as soon as possible, the government had to pay more attention to the growth of the GDP and of the national fiscal revenue, so it was impossible to give priority to promoting the independent innovation of enterprises. Fifth, as to the system of innovation support of the whole society, the foundation of Chinese industrial technology was still very weak and old-fashioned, and it was difficult to support the leading enterprises' independent innovation. It was very good on the whole that the former research institutes affiliated to ministries turned into enterprises, but their backbone strength were neglected to rely on to build a generic industrial technology R&D system through a new mechanism. The transformation of a state-owned specialized bank to a commercial bank was right, but a mechanism was neglected to build in order to solve the persistent problem that the financial system feared the independent innovation of enterprises, but liked enterprises to introduce and imitate.

"System structure characteristics determine the output capacity and the output characteristics of the system." When we implement the strategy of independent innovation and build an innovation-oriented country, the key is to establish the national innovation system oriented towards independent innovation. In general, the transformation from an NIS oriented towards imitative innovation to one oriented towards independent innovation needs four conditions. The first is that the enterprises must have the urge to carry out independent innovation. The second is that the government must be far-sighted in order to actively support and encourage the independent innovation of enterprises. The third is that the level of the economy and S&T must have developed to a certain stage. Especially, financial investment in basic research must be able to sufficiently support the knowledge innovation in universities and institutes; the R&D investment of the whole society must be enough to support the whole innovation of large enterprises and the technology integration facing mass production of SMEs. The fourth is that there must be a social culture, especially an organizational culture that encourages trial- and- error and innovation.

After 2006, China began to establish an NIS oriented towards independent innovation, which means that the internal structure and the operation mechanism of this system need to help produce more independent innovations in the economy. The report of the 17th CPC National Congress clearly pointed out that China would conscientiously implement the national medium- and long-term S&T development plan; adhere to the guidance policy of "independent innovation, stepping over key points, supporting development, leading the future" in order to cultivate and form strong ability for original scientific innovation; grasp opportunities and respond calmly to various problems in the progress of S&T and in the S&T revolution; be equipped with powerful capability for core technological innovation and occupy an active position in the fierce international economic competition in S&T; acquire great ability for system integrated innovation and the capability to import, implement, adopt, and re-innovate, and effectively absorb international innovation resources in an open environment; establish an efficient and fluent technology transfer mechanism, and a scientific knowledge diffusion mechanism to benefit the society both economically and socially; construct and shape a socialistic legal system with Chinese characteristics, an advanced innovation culture, a good social innovation and entrepreneur atmosphere, an innovation system and national education systems full of vigor and vitality in order to make innovation wisdom burst forth eagerly and allow a large number of innovative talents to emerge and then shape strong ability for independent innovation with the goal of supporting Chinese economic and social development and of becoming an innovative country in 2020.

These statements primarily reflect the basic idea of the construction of the NIS oriented towards independent innovation.

(2) The basic framework and function of the NIS oriented towards independent innovation

First, the basic framework of the NIS oriented towards independent innovation with enterprises as the subjects. The NIS is the IS at the overall national level. The IS of industry and enterprises evolves along with the NIS. The construction of an independent innovation-oriented IS needs the interaction, mutual support, and joint efforts of the government, enterprises, universities, scientific research institutes, and all of the social sectors. In this complex system, enterprises should be the subject of independent innovation, the government be the leader and supporter, and universities and research institutes be the providers of new technology, new talents, and new knowledge. All kinds of social intermediary organizations should endorse independent innovation (shown in Figure 6-5).

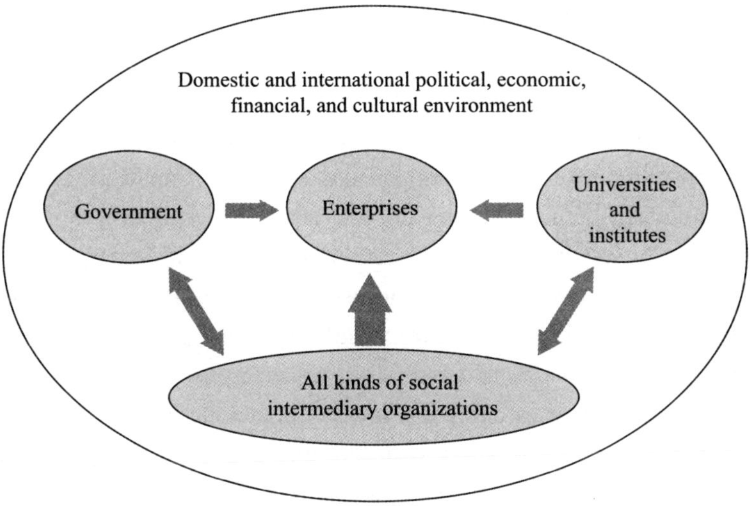

Figure 6-5 The prototype framework of an NIS

Second, the function of government in an independent innovation-oriented IS. The government plays a vital role in whether a country can produce more independent innovation. If the government plays a greater role in inducing and promoting independent innovation, it must deal with the following two questions perfectly: The first is that the government needs to invest funds properly in the fields of national defense, public welfare, and generic technology and in the fields where the industrial circle is not able to carry out R&D; the government must also support related R&D work. The second is that the government should focus on optimizing the innovation environment. The government should help enterprises create an independent innovation environment with appropriate policies and financial support. At the same time, the government cannot directly intervene in the operation and innovation activities of enterprises because the enterprises must become the actual subjects of innovation.

Third, the function of universities and scientific research institutes in the independent innovation-oriented IS. Universities and research institutes should provide new technology, new talents, and new knowledge needed for independent innovation. The first is that they can provide new knowledge and new technology required by enterprises, especially the bottleneck technology affecting the development of enterprises and industries, by transferring existing technology to them, accepting the responsibility entrusted to them for R&D and by establishing R&D centers together with enterprises. The second is that universities can adjust their teaching and educational methods as well as their major settings and provide personnel training for enterprises that cultivates professional talents required by enterprises in order to carry out independent innovation. The third is that universities and research institutes are often the information center of S&T development in related fields. Based on this, it is necessary for them to provide relevant information to serve the enterprises' independent innovation.

Fourth, the function of cultural inheritance in an independent innovation-oriented IS. Innovation is the soul of national progress. Without the culture for innovation, a nation is without hope. Therefore, we also need to create a cultural atmosphere conducive to independent innovation by the influence of the S&T development, the guidance of government policies, and the efforts of the whole society. The first step is to stimulate and develop innovation consciousness in children. From basic education to higher education, innovation education should be the main emphasis in education and teaching. The second step is that the government, enterprises, schools, institutes, families, and all kinds of social organizations should actively encourage people to innovate, while at the same time to tolerate adventure and failure and to reward innovators in various forms.

(3) The key problems in the construction of an NIS oriented towards independent innovation in the current stage

At present, China is in a crucial transition period of the NIS. In order to achieve this transformation, at least three problems should be solved.

The first problem is to cultivate the independent innovation consciousness of universities, scientific research institutes, and enterprises. Higher education should first carry out independent innovation and become the cultivator of talents for independent innovation. National research institutes and academic universities should be more than a subject of knowledge innovation, and not establish enterprises without competition advantages. Large enterprises should be the subject of technical innovation during the whole process, SMEs should be more than a subject of technology integration for the development of new products and for the mass production and marketing.

The second problem is the adjustment of the national innovation mechanism. The first thing is to remember Smith's thought concerning the division of labor to guide universities, institutes, and enterprises towards a reasonable division and cooperation. At the same time, the "incubator of technology commercialization" should be developed effectively to link up the technical supply of universities and institutes with the technical demand of enterprises. The second thing is to straighten out the innovation investment mechanism, and increase the R&D investment of the whole society, especially governmental investment in basic research. The third thing is to straighten out the social incentive mechanism of innovation and strengthen and normalize the market mechanism as to its suitability to select the superior and eliminate the inferior. The fourth thing is to improve the government's policy-guiding mechanism and give a strong degree of "ex post incentive" to those who reach outstanding achievements in independent innovation. It would be a good idea to give the independent innovation enterprises a tax preference for value-added tax. The fifth thing is to establish and improve the credit mechanism for science and technology.

The third problem is to straighten out and strengthen the social supporting system of independent innovation. The technological foundation of key industries should be profoundly improved. The backbone of strength from the research institutes that were formerly affiliated to ministries and have transformed the system should be chosen in an attempt to construct an R&D system of generic industrial technology via a new mechanism. State-holding commercial banks should be encouraged to support the independent innovation of enterprises by their policies, especially the independent innovation of large enterprises at the fore-end of the industrial innovation chain, and SMEs that undertake "technology integration facing mass production."

6.3 The establishment of a sectoral innovation system (SIS) oriented towards independent innovation

6.3.1 The connotations of an SIS

Breschi & Malerba (1997) put forward the concept of sectoral innovation system (SIS) based on the NIS and technology system research. They think it is “a set of heterogeneous agents carrying out market and non-market interactions for the generation, adoption, and use of technologies and for the creation, production, and use of products that pertain to a sector.”

Zhang Feng & He Chuanqi (1999) think that SIS refers to a network system made up of knowledge and technological innovation institutions and organizations related to the sectors, including both domestic and international ones; their activities and behaviors promote and improve the innovation ability and competitive ability of the sectors. Liu Xielin (2000) claims that the SIS is the confirmation of a network relationship; the nodes are enterprises, knowledge production institutions, intermediary agencies, and customers in the chain of production. The interaction among the nodes is mainly reflected in that of enterprises and departments classified according to technology and network connection characteristics. The connection forms among nodes are trade and knowledge flow. The innovation opportunities of the SIS are closely related to the number of nodes and the degree of interaction among them. The more nodes are, the more frequent the knowledge or trade flow is, and the more innovation opportunities are.

Zhang Naiping (2003) believes that the subjects of sectoral innovation are enterprise groups and the government, and that sectoral innovation is based on sectoral technological innovation with the characteristics of integrity, fundamentality, systematicness, and integration; all aspects of sectoral innovation jointly promote sectoral competitiveness. Zhang Zhihe (2003) thinks that the SIS is a network system, which takes market demand as the driving force, policy regulation as guidance, good domestic and international environment as security, innovative technology supply as the core, and the realization of specific sectoral innovation as the goal.

Combining the domestic and foreign scholars' definitions of SIS, we can draw the following conclusion: SIS is a systematic concept, which refers to the network system formed by the elements in each node in the sectoral chain (such as enterprises, research institutes, universities, intermediary agencies, customers, etc.) in the process of the dissemination and diffusion of knowledge and technology. Its function is to accelerate a sectoral technology breakthrough, especially that of generic sectoral technology, and to promote the sustainable development of the sector.

6.3.2 The framework of an SIS oriented towards independent innovation

(1) A brief review of the elements and components of an SIS

The SIS is a complex system. Breschi & Malerba (1997) think it includes a knowledge and a technological base, and an entry and exit mechanism. The agents composing the sectoral system can be individuals and organizations. These organizations may be companies, universities, financial institutions, government agents, trade unions, and technology associations; these individuals can be consumers, entrepreneurs, and scientists. They interact through processes of communication, exchange, cooperation, and competition.

The research of McKelvey & Orsenigo (2001) on the European pharmaceutical industry finds that innovation subjects include each company in the industrial chain, public or private research institutes, research institutes of universities, financial institutions, governmental supervision agents, and consumers. Every innovation subject does not have a simple mutualistic relationship, but a complex relationship of competition, cooperation, upstream and downstream supply, and R&D provision. Zhang Feng & He Chuanqi (1999) think that an SIS takes enterprises as the core and nationwide scientific institutions, education institutions, and intermediary agencies as the backing in order to shape an open innovation network. In this network, the innovation resources of knowledge, technology, information, talents, and capital can flow quickly. Zhang Naiping (2003) believes that an SIS is a technology and economy integrated system, which is organically made up of sectoral technological innovation, sectoral organization innovation, and sectoral management innovation. The SIS is affected by national resources, governmental policies, market conditions, the international economic and technological environment, enterprise behavior, the running conditions of each subsystem, etc. There are energy exchange and feedback mechanisms between the subsystems and the SIS. The SIS model is shown in Figure 6-6.

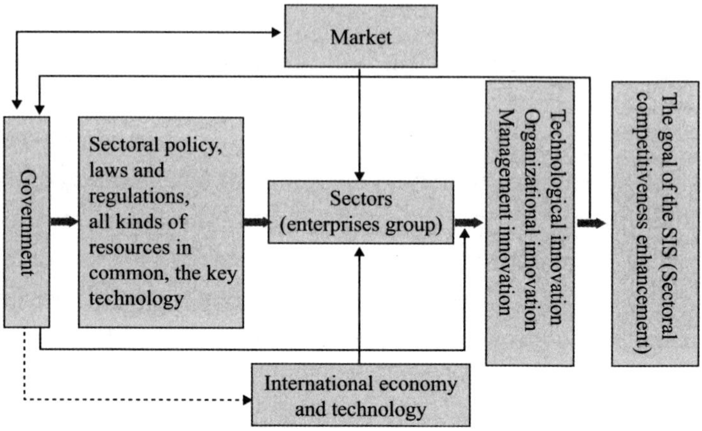

Figure 6-6 The operation model and mechanism of an SIS

Zhang Zhihe (2003) points out that the essence of sectoral innovation is the re-combination of production elements and the introduction of a production system. The elements of an SIS include subject elements, environmental elements, and connecting elements. The function of an SIS is to realize the organic combination of every element and the organic linkage among the various elements with the use of appropriate tools, especially policy tools. The model of an SIS built by Zhang Zhihe (2003) is made up principally of a policy system, a technology system, an evaluation system, the background environment and a connecting mechanism. A specific model of SIS is shown in Figure 6-7.

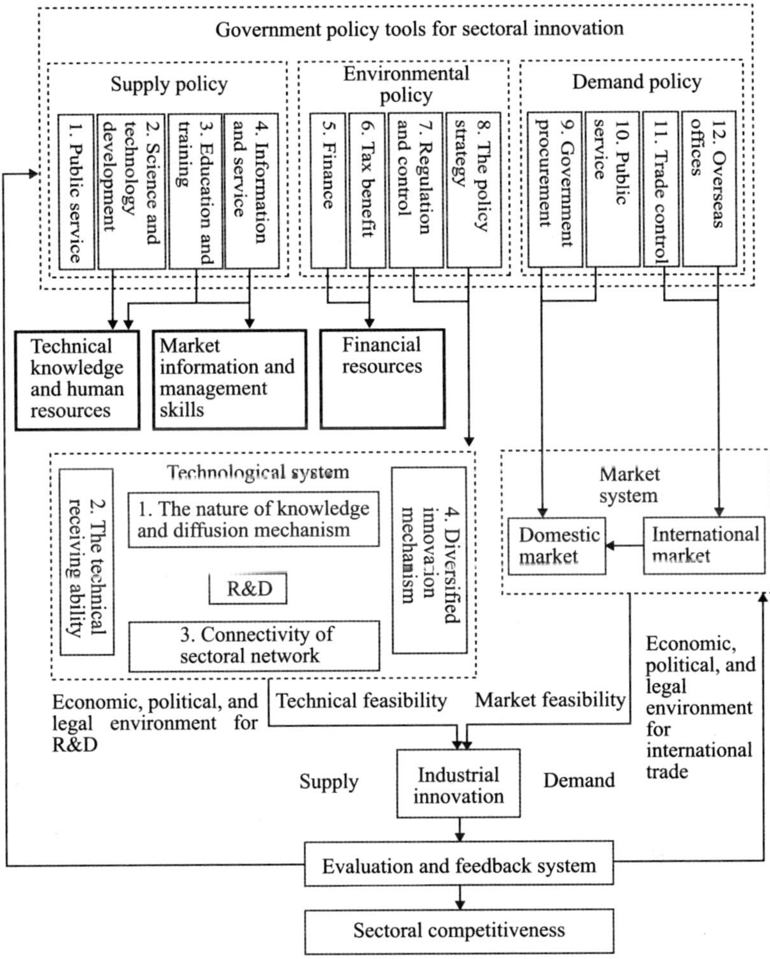

Figure 6-7 An SIS model built by Zhang Zhihe (2003)

Li Chunyan & Liu Lizhen (2007) constructs a model of SIS with technological innovation as the core. She thinks that inside a sector, the innovation process starts with various types of technical innovations to meet the technological demands of enterprises, such as incremental innovation, basic innovation, the reform of new technological systems and technological economic modes, etc., followed by the product innovation. In this stage, the achievements of technology development will be trial-produced into products, which will be commercialized and functionalized through further processing and improvement in order to be ultimately accepted by the market. Finally, there is market innovation stage. At this stage, new technologies and products spread along various channels; the production of industrial clusters of new products begins to form and the market will be greatly developed. This forms industrial innovation in the end. She constructs an SIS model with technological innovation as the core as shown in Figure 6-8.

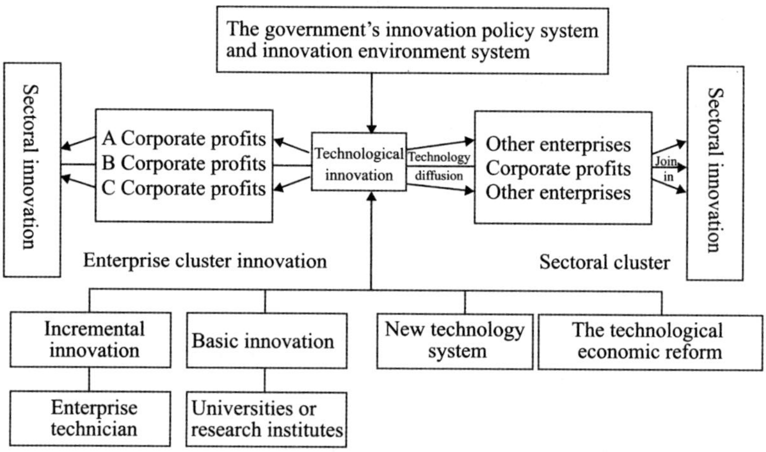

Figure 6-8 An SIS model built by Li Chunyan & Liu Lizhen

By analyzing domestic and foreign scholars' discussions on the model of SIS, we can find that they have proposed some models of SIS with different characteristics based on their own understanding and their own analysis of the structure of an SIS. In particular, the subject elements, environmental elements, links among the units, and comprehensive target are indispensable elements for the construction of a model of SIS innovation system.

(2) The elements of an SIS oriented towards independent innovation

On the whole, the elements of an SIS oriented towards independent innovation are shown in Figure 6-9.

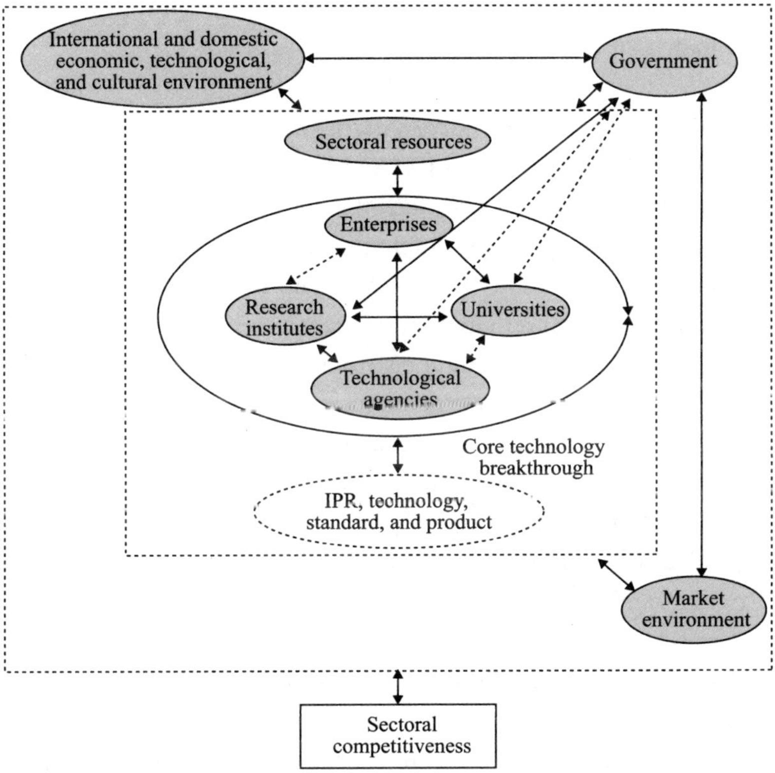

Figure 6-9 The elements of an SIS oriented towards independent innovation

First, innovation subjects. The subjects of sectoral innovation are the direct participants in innovation activities and are responsible for main innovation activities, and they may be enterprises, universities, and scientific research institutes. There are no essential differences between the characteristics and functions of the subjects of sectoral innovation and those of an NIS.

Second, supporting elements. They can be the intermediary S&T agent and financial institutions, which are not directly involved in sectoral innovation, but serve the innovation subject by supportive activities, and then influence the implementation, diffusion, and dissemination of innovation activities. At certain times, these elements can even determine whether innovation subjects can implement an innovation.

Third, innovation environment. This factor affects the behavior of an innovation subject and the effectiveness of supporting elements, and it also influences the efficiency of an SIS. The innovation environment includes governmental policies and social culture, the basic domestic conditions of economy and technology, the domestic and international market environment, etc. The innovation function of the government and the cultural environment has been described in the preceding part. Market environment can be divided into commodity market environment and element market environment. It is the platform where enterprises can obtain information, access resources, create value, compete with others, and acquire profit, and it is the medium that links the production enterprises with upstream suppliers and downstream distributors and customers.

Fourth, links among the units. These links are key factors influencing the operating efficiency of an SIS, including legal and policy links, financial flows, informational flows, S&T flows, and trade chain (Niosi et al., 1993). Connecting elements can be intuitively shown as follows: i) Legal and policy links. Operation behaviors of enterprises are mainly regulated by the implementation of relevant laws and policies and then a market transaction environment conducive to innovation will be created. ii) Financial flows. This refers to the connection among subjects of innovation and between innovation subjects and the supporting elements based on capital relationships. For example, financial institutions provide subjects of innovation with required funds; cross shareholding among enterprises is also a means of connection. iii) Informational flows. It means the information interaction among all the types of entrepreneurship subjects, supporting elements, and environmental elements, and the information interaction usually involves the delivery of product, personnel, technology, market, finance, and policies. iv) S&T flows. That is the connection of S&T resources and the active link that occurs among innovation subjects; it directly shows the transfer and diffusion of S&T knowledge and information. For example, the cooperative innovation of subjects is one way that the active link of S&T resources occurs. v) Trade chain. It means the flow of goods and services in each node of sectoral chain. It is one of the most basic forms of link in an SIS.

6.4 The construction of an enterprise innovation system (EIS) oriented towards independent innovation

6.4.1 The connotations of an EIS

Improving the enterprise innovation system (EIS) is an effective guarantee for the promotion of the ability of enterprise innovation. Xu Qingrui (2000) points out that the construction of an enterprise technological innovation system has to strengthen the relationship among the R&D department, the production department, and the marketing department based on the entrepreneurship of the enterprise leaders, the R&D system and the education and training system, capital, management tools, and enterprise culture and system. Li Xiangyin & Wang Yapin (2002) think that an EIS is made up of enterprise innovative information system, power system, innovation search and selection system, innovation process system, innovation method system, innovation ability system, innovation support system, and the relationship among these systems and between these systems and the external environment.

Wang Bin (2004) defines EIS as a system shaped by innovation elements with different status in a certain area. He thinks that an EIS is the sum of all the conditions, rules, structures, processes, and ways of carrying out activities in which the innovators of enterprises take advantage of technological inventions, management discovery, institutional change, and market opportunities, as well as by reforming the essential production factors and conditions along with the methods of allocating resources to gain commercial success of the reform. Cao Longcan (2006) claims that enterprise independent innovation system refers to the network, the organization, and the system established by the enterprise so as to achieve key core technology breakthroughs and acquire independent IPR.

Gu Qiuqin (2008) divides an EIS into two subsystems: The enterprise technological innovation system and the enterprise institutional innovation system (property rights system, incentive system, and daily management system). She thinks that enterprise technological innovation system is an organic whole shaped by the connection and interaction of the structure of the enterprises' technological innovation organization, the innovation resource elements (personnel, funds, equipments and so on), the innovation mechanism, and the institutional and external environments under the support of the enterprise technological innovation system. Its goal is to promote the enterprise's capability for technological innovation and competitiveness.

Some scholars define EIS as an "internal network system which is beneficial to the enterprise for creating, introducing, improving, and popularizing new knowledge and technology more effectively."

Analyzing these scholars' definitions of EIS comprehensively, we could come to the conclusion that an EIS is a network formed by the mutual link and interaction existing within the enterprises, among the enterprises, and between the enterprises and the subjects of the external environment.

6.4.2 The framework of an EIS oriented towards independent innovation

(1) A brief review of the constitution and elements of an EIS

Guo Bin et al. (1997) believe that there are six factors in a complete enterprise technological innovation system: Entrepreneur and entrepreneurship, strategy, organization structure, R&D, capital, and talents. The first three factors determine the effectiveness and the efficiency of enterprise innovation activities, and the last three are enterprise innovation resources.

Xia Baohua (2001) thinks that a complete enterprise technological innovation system is mainly made up of enterprise production-learning system, enterprise technological innovation search system, enterprise technological innovation exploration system, and enterprise cultural innovation system. Wang Bin (2004) believes that the subjects of an EIS includes the enterprise, the entrepreneur, and the intrapreneur; he also believes that the objects include product innovation and process innovation. Gu Qiuqin (2008) claims that the enterprise technological innovation system is made up of the subject, the mechanism, and the environment.

Some scholars think that an EIS is a systematical whole which is made up of factors as follows: People with initiative spirit and innovation goal, innovation institution, innovation motivation, innovation skills, innovation means, innovation material, innovation capital, innovation information, innovation environment, and innovation control. Some scholars also think that an EIS is a comprehensive system that consists of technological innovation, organizational innovation, managerial innovation, and system innovation.

Combining the above analysis, we can see that the constitution of an EIS is not only confined to the subjects within and among enterprises, but also has a close relationship with the external environment of the enterprises; the elements of the subject of innovation, the supporting elements of innovation, the environmental elements of innovation, and the elements of innovation connection are the basic units of an EIS.

(2) The constitution of an EIS oriented towards independent innovation

Calantone et al. (2010) think that looking at it from inside the enterprise, independent innovation involves innovative idea, technological innovation, engineering design, sample development, middle test, pilot production, product sales, after-sale service, and the realization of commercial profit. The internal factors that influence enterprise independent innovation include R&D organization institutions, innovation resource elements, supporting and safeguarding mechanism, and innovation process operation mechanism. Looking at it from outside the enterprise, independent innovation will be affected by external economic, technical, political, cultural, and other factors, especially governmental policy, customer behavior, and market environment.

To realize a more independent innovation, enterprises need to establish an independent innovation system at the enterprise level. Calantone et al. (2010) believe that the enterprise independent innovation system is an entity shaped by the mutual connection and restraint, and the interaction of the factors whose core is the innovation process that enterprises "start via innovative ideas and R&D, and terminate with market realization." Calantone et al. (2010) consider the goal to be achieving a key technology breakthrough, acquiring independent IPR, and enhancing the enterprise's ability for innovation and competitiveness.

Cao Longcan (2006) thinks that, from the point of view of the source of technology and the innovation strategy, the enterprise independent innovation system includes the endogenous innovation subsystem in the competition stage, and the cooperation innovation subsystem as well as the introduction innovation subsystem established on the basis of the endogenous innovation subsystem. This would constitute the core part of the enterprise independent innovation. The environmental subsystem is also an important supporting system of enterprise independent innovation, which plays a significant role in supporting and guaranteeing the enterprises to achieve independent innovation.

To realize the transformation from imitative innovation to independent innovation, it is more necessary for enterprises to construct an EIS oriented towards independent innovation. Such construction of an EIS emphasizes not only innovation independence, but also the completeness and systematicness of an innovation system which at least includes an innovation subject subsystem, a supporting subsystem, a subsystem that connects the elements, and an environmental subsystem.

First, the innovation subject subsystem. The enterprise is the subject of independent innovation. This means that enterprises should be the subject of not only innovation input and output, but also of innovation earnings. Entrepreneurs and intrapreneurs play a key role in an EIS oriented towards independent innovation. The entrepreneur is the core, and entrepreneurship is the key factor which determines whether an enterprise is able to carry out independent innovation. Drucker, the founder of modern management, thinks that the most important thing about entrepreneurship is innovation; the entrepreneur is the initiator and the organizer of the independent innovation of enterprises. Intrapreneurs are indispensable in the independent innovation-oriented EIS. The so-called intrapreneurs are the people engaged in the work similar to that of entrepreneurs. Their pursuit of creative or new solutions to challenges confronting the firm, including the development or enhancement of old and new products and services, markets, administrative techniques, and technologies for performing organizational functions, as well as changes in strategy, organizing, and dealings with competitors are innovations in the broadest sense. (Hisrich & Brush, 1986; Antoncic & Hisrich, 2000, 2001).

Second, the innovation supporting subsystem. To have more independent innovation, enterprises also need corresponding supporting conditions. Therefore, an innovation supporting subsystem is an indispensable element in an EIS oriented towards independent innovation, including universities, scientific research institutes, S&T agencies, financial institutions, etc. In particular, universities and scientific research institutes are the sources of knowledge and technology for enterprise independent innovation, and S&T agencies provide a bridge that connects enterprises with other supporting elements.

Third, the element connection subsystem. The links in an EIS can be divided into an external link chain and an internal one. The external link chain reflects the relationship between enterprises and innovation supporting elements and between enterprises and environmental elements, involving financial chain, S&T chain, informational chain, legal and policy chain, social chain, and trade chain. The internal link in the chain mainly refers to the internal mechanism of enterprise innovation. It is the key to promoting sustainable innovation and an efficient operation of enterprises. Specifically, the mechanism of an EIS mainly includes enterprise innovation operation mechanism, enterprise innovation incentive mechanism, and enterprise innovation development mechanism. The internal mechanism of enterprise innovation is established by the enterprise to guarantee the effective operation of a series of innovation activities from the conception to the product commercialization, including the organization, management system, and operation procedures of innovation. The establishment of the internal mechanism of enterprise innovation is helpful for an enterprise to effectively allocate innovation resources, to coordinate organization during the innovation process, and to promote a smooth innovation process.

Fourth, the innovation environmental subsystem. To have more independent innovation, enterprises also need corresponding environmental conditions, so in an EIS oriented towards independent innovation, an innovation environmental subsystem is also indispensable, including the government policy, the market environment, the international economic and technical environment, as well as market organization. Among them, the government should be the guide for enterprise independent innovation; other environmental providers influence enterprise independent innovation as the partners or competitors of the enterprise.

The mode of an independent innovation-oriented EIS is shown in Figure 6-10.

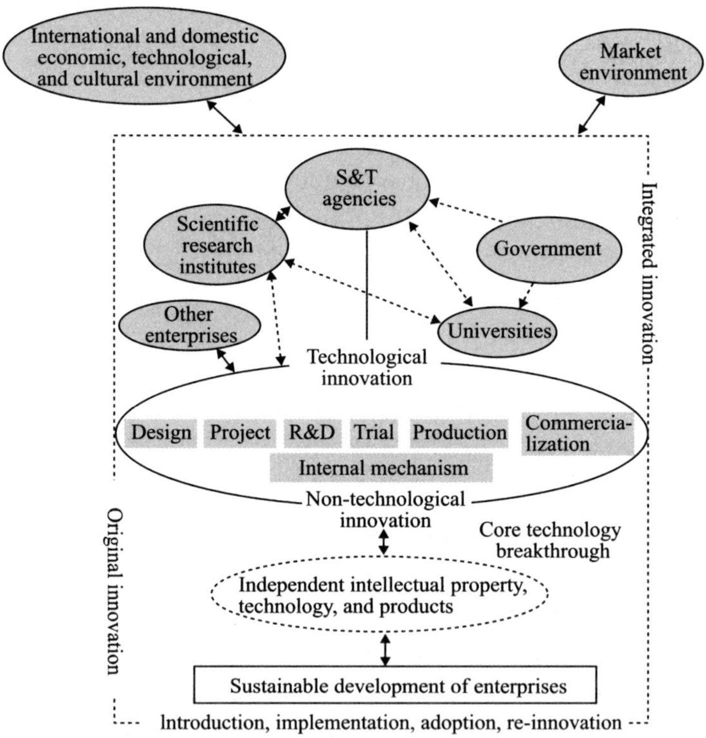

Figure 6-10 The model of an independent innovation-oriented EIS

References

Antoncic, B. & Hisrich, R.D. Intrapreneurship modeling in transition economies: A comparison of Slovenia and the United States. *Journal of Developmental Entrepreneurship*, 2000, 5 (1): 21-40.

Antoncic, B. & Hisrich, R.D. Intrapreneurship: Construct refinement and cross-cultural validation. *Journal of Business Venturing*, 2001, 16 (5): 495-527.

Breschi, S. & Malerba, F. Sectoral innovation systems: Technological regimes, schumpeterian dynamics, and spatial boundaries. In Edquist, C. (ed.) *Systems of Innovation*. London: Frances Pinter, 1997: 130-156.

Calantone, R.J., Harmancioglu, N. & Droge, C. Inconclusive innovation "returns": A meta-analysis of research on innovation in new product development. *Product Innovation Management*, 2010, 27 (7): 1065-1081.

Cao, L. Research on enterprise independent innovation system and model. Doctoral dissertation of Hefei University of Technology, 2006.

Fei, Z. *The Cambridge History of China* (Vol. 2: Republican China). Beijing: China Social Sciences Press, 2006.

Freeman, C. *Technology Policy and Economic Performance*. London: Pinter, 1987.

Gu, Q. The main body composition and synergy of enterprise technology innovation. *Journal of Southwest University for Nationalities*, 2008, 29 (7): 246-250.

Guo, B., Chen, J. & Mao, Y. Discussion of technological innovation system and mechanism of stated-own large and medium-sized enterprises. *Journal of Industrial Engineering and Engineering Management*, 1997, 11 (6): 59-65.

Hisrich, R. D. & Brush, C. Characteristics of the minority entrepreneur. *Journal of Small Business Management*, 1986, 24: 86-99.

Li, C. & Liu, L. Industrial innovation system production mechanism and structural model. *Science of Science and Management of S.&T.*, 2007 (1): 50-55.

Li, X. & Wang, Y. Discussion of enterprise innovation system. *Science*

and Technology Management Research, 2002 (1): 78-80.
Liu, X. *Chinese Technology Innovation System in the 21st Century*. Beijing: Peking University Press, 2000.
Lundvall, B.-A. *National Systems of Innovation: Towards a Theory of Innovation and Interactive Learning*. London: Pinter Publishers, 1992.
McKelvey, M. & Orsengio, L. Pharmaceuticals as a sectoral system of Innovation. Paper presented at SPRU workshop and delivered to EU as part of the EU project "European Sectoral Systems of Innovation", 2001.
Merriam-Webster. *Merriam-Webster's Collegiate Dictionary* (11th Edition). Springfield, MA: Merriam-Webster, Inc., 1995.
Niosi, J., Saviotti, P., Bellon, B. & Crow, M. National systems of innovation: In search of a workable concept. *Technology in Society*, 1993, 15 (2): 207-227.
OECD. The knowledge-based economy. *OECD Publication*, 1996.
Qu, T. *China University Education Development History*. Beijing: Peking University Press, 2006.
Wang, B. Research on enterprise innovation system. Doctoral dissertation of Sichuan University, 2004.
Xia, B. *Structure of the Sustained Technological Innovation of Enterprises*. Shenyang: Northeastern University Press, 2001.
Xu, Q. *The Management of Research, Development and Technology Innovation*. Beijing: Higher Education Press, 2000.
Zhang, F. & He, C. *The National Innovation System—Engine of the Second Modernization*. Beijing: Higher Education Press, 1999.
Zhang, N. China's auto industry innovation system. Doctoral dissertation of Wuhan University of Technology, 2003.
Zhang, Z. Industry innovation system of the Optics Valley of China. Doctoral dissertation of Wuhan University of Technology, 2003.

Appendix

Enterprise Case 1

Huawei—Independent innovation as the challenger's passport

1. The progress of Huawei's independent innovation

Huawei Technologies Co., Ltd. (Huawei in short) was founded in 1987. It is an employee-owned high-tech private enterprise and also a leading global information and communication solution provider. Huawei is mainly engaged in the research, development, production, and sales of communication network technology and products. It specifically provides telecom operators with solutions for optical networking, fixed broadband, mobile broadband, and value-added services. Huawei's products and solutions cover mobile networks (LTE/HSPA/WCDMA/EDGE/GPRS/GSM, CDMA2000 1xEV-DO/CDMA2000 1X, TD-SCDMA, and WiMAX), core networks (IMS, Mobile Softswitch, NGN), networks (FTTx, xDSL, optical networking, routers, and LAN switches), value-added telecom services (IN, mobile data services, BOSS), and terminals (UMTS/CDMA). In 2010, Huawei's global revenue reached $28 billion with a year-on-year growth of 24.2%. In 2014, Huawei recorded a profit of $5.5 billion. Its products and solutions serve more than 140 countries, including 45 out of the top 50 global countries and 1/3 of the world's population. This establishes its position as the second largest information and communications technology (ICT) provider in the world, and it ranks No. 351 in the *Fortune* Global 500 list and No. 18 among the world's most innovative companies. Huawei has initially developed into a global company with strong competitiveness through independent innovation.

1.1 Self-developed large-scale switches—The beginning of innovation

The period from 1987 to 1995 is the initial stage of Huawei's development. Huawei insisted on product management and concentrated its resources on successful factors; moreover, it improved its technical strength, market positioning, and management capabilities with its local advantages. It focused on imitative innovation and the rural market. In 1987, Huawei was only a sales agent of private branch exchange (PBX) switches. After about two years' development, Huawei developed its marketing network and team and grasped the pulse of the domestic telecommunications industry and the communications market. In 1990, Huawei started to research and develop the PBX targeting hotels and small enterprises independently and then released the PBX BH301. In 1991, it released the 512-type portal HJD48 and the vertical and horizontal type conversion machine JK1000. In early 1993, Huawei launched its large-scale equipment C&C08, which indicated that Huawei had accumulated independent technology and entered a new stage. Its C&C08 machine has a unique technology and is the first to use optical fiber as the connecting material, which is able to meet the special needs of lightning protection, power consumption, and remote modules. Switches in other countries were mostly connected by a cable, the greatest weaknesses of which are its high technical maintenance requirement and the high cost of laying it in places where users are decentralized. Thus, it is not suitable for a remote market. However, optical fiber proves to be the most suitable for such kind of market. Huawei's fiber-optic technology was especially adapted to the needs of the rural market. Under the condition that large-scale switches in other countries were connected via cable, Huawei developed the machine connected to optical fiber, which had a large potential market. This technology gave Huawei significant power to transform its development from rural areas to urban areas.

1.2 The development and upgrading of independent innovation

Before 1995, Huawei's market was concentrated in rural areas. Since 1995, Huawei has started to carry out an internationalization strategy. It made its way to the international market through independent innovation, and gradually formed a strong marketing and service network. Regarding innovation, it changed from imitative innovation to cooperative innovation, then to independent research. Its management innovation, system innovation, and market innovation gradually kept up the pace of technological innovation to promote the growth of the enterprise.

First, Huawei has achieved a breakthrough by buying patented technology. As a percentage of sales, Huawei's annual R&D expenses is more than 10%. What Huawei has done is not only improve the function and promote the integrated ability of the achievements it has bought from multinational companies, but also make technological progress in the engineering design and realization. For the core technology it lacked, Huawei gets access to the international market by buying patents and paying the patent licensing fees. Meanwhile, it gradually survives in the competitive market by reducing its costs.

Second, Huawei has achieved innovative development through cooperative innovation. On the one hand, Huawei cooperates with universities and scientific institutes. It regularly allocates funds to establish an ongoing relationship with universities and research institutes and makes full use of their information platform and channel, to grasp the domestic and international cutting-edge technology trend and industry development situation in the field of electronic information. On the other hand, Huawei and some large companies with powerful technical strength form a strategic alliance to jointly carry out technological innovation. On the technical side, Huawei has established a joint laboratory with T1, Motorola, Intel®, Agere, Altera, Sun, Microsoft, NEC, and other world-class enterprises. It has established Yumeng Company with Panasonic and NEC, and has established Huawei 3COM Company (H3C) and a 3G development laboratory with 3COM, and has been cooperating with others to drive the industrial value chain.

Third, Huawei has constructed an innovation concept based on customer demand. Its technological innovation is oriented towards customer demand. They produce whatever customers need. Things that can be sold out and products that are pioneering the market are the real technological demands of customers. An enterprise must make a profit in order to develop, and profits come from customers, so whether it is core technology or appearance design, customer orientation is the most fundamental policy. Only by meeting the needs of users, can technological commercialization be truly achieved.

1.3 Management innovation aligning Huawei with international standards

At the early stage, the management level of Huawei lagged behind market development. In order to become bigger and stronger, Huawei had to improve its management and technological innovation. At the same time, it had to attach importance to management innovation and system innovation in order to promote comprehensive development.

In 1995, the formulation of "the basic law of Huawei" started its management innovation. It helped Huawei make a change from rule by man to rule by law, which laid a solid foundation for its subsequent stable development. In 1997, Huawei started to systematically introduce the management mode of world-class management consulting companies and established its management system based on IT, in line with international standards. As to changes in business process, with the assistance of the project to reform financial and human resources (Hay), Huawei took market management, integrated product development (IPD), integrated supply chain (ISC), and customer relationship management (CRM) as its focal point in comprehensively carrying out its reform. Huawei introduced the best practice of this industry, and built an entire IT infrastructure to support its operation. Concerning financial management, Huawei launched the integrated finance servers-services (IFS) project to improve the capacity of its management operation. As for the supply chain, Huawei implemented the reform of an ISC, strengthened the performance management of purchasing, and boosted the supplier certification process based on the best practice in this industry, TQRDCE—Technology, Quality, Responsiveness, Delivery, Cost, and Environment. After more than five years of management improvement and the implementation of a customer demand-driven development process and supply chain process, Huawei had gained differentiation competitive advantages in accordance with consumer interest, and had further consolidated its core competitiveness.

1.4 The sustainable development of Huawei

In 2003, Huawei launched a full set of R4 commercial products of 3G as the first one in the world. Moreover, it realized the world's first commercial network based on R4 architecture in the United Arab Emirates, and completed the first R4 device communication and docking among different manufacturers in the world. On the B plane of the first-degree long-distance tandem network of China Mobile Communications Corporation (CMCC), Huawei realized the commercialization of the world's largest mobile soft switch tandem network, and the resource usage rate of the IP bearing the network at the peak reached 40%.

In 2004, the 3G data communication card, developed successfully by Huawei, started to supply the operators from Hong Kong of China and Europe, and Huawei became one of the only two manufacturers in the world. In 2005, during the 3GSM World Congress in Cannes, France, Huawei released the UMTS distributed base station solution, which was innovative and could support the rapid building of a network. In 2006, regarding the expanding market, H3C realized the depth extension of the IP network development from layers No. 1–3, to layers No. 4–7, and Huawei devoted itself to the transition from a network equipment supplier to a provider of complete business solutions.

In 2007, based on the Qualcomm test terminal MSM6280, Huawei successfully completed a video phone based on the UMTS900 and the high-speed downloading with HSDPA3.6 Mbps as a part of a current European network experiment, and completed the IOT test of the terminal and system of the UMTS900 in Shanghai. In 2008, Huawei was chosen as one of the top 10 "World's Most Influential Companies" by *Business Week*. The UMTS/HSPA networks in North America were used commercially on a large scale for the first time; the next generation wireless network was also built for Canadian operators Telus and Bell. In 2009, Huawei first released the end-to-end 100G solution from the router to transmission system, and successfully delivered the world's first LTE/EPC commercial network. The number of LTE's commercial contracts awarded to Huawei ranked the first in the world, and the market share of wireless accessing ranked the second in the world. By 2010, Huawei had deployed more than 80 SingleRAN commercial networks worldwide, 28 of which had commercially released or would release LTE/EPC business soon. Through independent innovation, Huawei had developed from a private enterprise which engaged in agency business to a large multinational S&T company with an extensive and powerful ability for innovation.

2. The four magic keys to Huawei's independent innovation

Huawei's development is inseparable from the policy support of Shenzhen government, and also Huawei's independent innovation strategy.

2.1 The policy support from Shenzhen municipal government is a strong guarantee for Huawei's independent innovation

Shenzhen government had set forth policies to improve the regional innovation system and to strengthen the protection of IPR, which promoted the development of S&T enterprises, such as Huawei, to a certain extent. In 2004, the "No.1 File" of Shenzhen declared to improve the regional innovation system and promote the development of the high-tech industry. In this way, Shenzhen government paved the way for enterprises to conduct research on core technology and basic patents in many aspects, like land, talents, IPR, etc. For the enterprises undertaking the road towards independent innovation, such as Huawei, Shenzhen government proposed a policy of "no intervene, but administration." "No intervene" means the government stayed completely out of the matter so that enterprises were able to adapt to the fierce competition in the market and were free to develop as necessary. "Administration" means the government should play a positive role in service and management. Besides, the government set up an intellectual property office to strengthen effective protection of intellectual property, including improving protection measures, reducing patent fee standards, and raising the penalties. In addition, the government used a "visible hand" in terms of finance and tax to vigorously support Huawei in providing a good living environment. Huawei's growth also tells us that without a good policy environment, an enterprise cannot achieve great success.

2.2 Core technology with independent IPR, the key factor of Huawei's successful independent innovation

Core technology with complete independent IPR is the basis of Huawei's foundation and the main reason for Huawei's strong ability to succeed in independent innovation and international competition. More than 10,000 professionals and technicians in Huawei participate in R&D each year. The proportion of R&D spending in sales revenue is more than 10%. For example, in the R&D of 3G, Huawei has invested 3,500 people and as high as 4 billion yuan. That Huawei spares no effort in R&D allows it to obtain a large number of core patents with independent IPR. At present, Huawei has the most patents in China, 85% of which belong to invention patents. The annual growth rate of the number of patent applications is more than 100%, and the number surpasses 1,000. Moreover, Huawei has won four National Science and Technology Progress Awards consecutively within three years. By Oct. 31, 2010, Huawei had applied for 31,869 Chinese patents, 8,892 PCT international patents, and 8,279 overseas patents; and the number of authorized patents had reached 17,765, including 3,060 overseas patents. In the field of LTE/EPC, the number of Huawei's basic patents has been leading the world. By the end of 2010, Huawei had joined 123 industry standard organizations around the world, such as 3GPP, IETF, ITU, OMA, NGMN, ETSI, IEEE, 3GPP2, etc., and had submitted more than 23,000 cases to those organizations. Besides, Huawei was a member of the board of directors of many authoritative organizations, like OMA, CCSA, ETSI, ATIS, and WiMAX forum, and had held more than 180 incumbent positions. Over 6,400 LTE/EPC cases had been proposed to 3GPP. Huawei's approved proposals on core standards of LTE wireless communication were nearly 20% of the total number worldwide. With its R&D strength and IPR, Huawei grasped the first round of commercial opportunities in the construction of the international market, and consolidated its international brand and market position. Based on its communication network technology, Huawei insisted on following the strategy of independent innovation, which was the key factor behind its growth.

2.3 Valuing and cherishing talents is the pillar that makes Huawei's independent innovation invincible

Huawei's development is derived from technology, which depends on talents. Huawei creates a "people-oriented" technological innovation system and an innovation incentive mechanism to cultivate innovation culture, and shapes effective industry-university-research technology alliances. The technical personnel's salary in Huawei is pretty good. The new employees can receive the same salary and welfare during their intern training as they get when actually working. If they hold internal stocks, it is also easy to cash out. Valuing and cherishing talents is the pillar on which Huawei's impregnable position depends both at home and abroad.

Huawei has set up 29 training centers in the world to train the technical personnel for local markets, and actively promotes the localization of its staff. The worldwide localization management not only deepens its understanding of local markets, but also contributes to the social and economic development for the nations and regions where Huawei branches are located. Huawei develops complete and comprehensive human resources management systems for its principles, working environment, staff training, career development, and the communication between its leadership and staff.

Huawei's business criteria require its employees to maintain a high level of professional ethics and standards of behavior. The criteria include manifesting sincerity, basic operating principles, and responsibility. Huawei's sincerity culture will make it become a great company that can fully realize the civic value of modern enterprises, strictly follow the morality of corporate citizenship, and faithfully follow the law in the host country.

Regarding the working environment, the departments responsible for employees' relationships should regularly investigate the working environment, such as by organizing environment surveys and customer satisfaction surveys, including workplace environmental monitoring mechanisms and annual physical examinations of its employees. Besides its internal review, Huawei has also introduced external authority (e.g. DNV) to inspect, in order to guarantee a better working environment for its employees.

In training, Huawei has made several efforts to turn training into a learning organization. An important measure is the official registration of Huawei University in 2005, which provides many training courses for its staff and customers, including cultural training and job training for new employees and training for customers.

Concerning career development, Huawei's qualification has a two-way promotion channel, combined with post demand, and this allows the personnel who have management skills and a potential managerial ability to smoothly grow into managers; it also makes it possible for those staff members with technological skills and advantages to become experts in some fields through their own efforts. Thus, this system provides its employees with excellent opportunities for career development.

Regarding salaries, Huawei not only follows local laws about the standard of minimum salary requirements, but also introduces a very competitive salary system. In order to remain invincible in market competition, Huawei's HR department has a long-term cooperation agreement with consulting companies such as Hay Group and Mercer, investigates salary data on a regular basis, and makes some adjustment for employees' salaries according to the survey and company performance.

As regards to communication, Huawei has an open communication channel between the company and its employees. Employees can express their own opinions and make suggestions to their direct supervisors, and they can also discuss their problems with higher-level leaders according to the open policy. Huawei's communication channels include, but are not limited to, the following: CEO mailbox, Open Day, employee relations specialists, and rationalization suggestion box. Huawei's relatively perfect measures for recruiting and retaining talents effectively promote its growth.

2.4 Managers' spirit of innovation, the source of Huawei's independent innovation

Entrepreneurs with a spirit of innovation are enterprises' key to overcoming limits and achieving independent innovation. Enterprise independent innovation requires managers to have a spirit of innovation and the willingness to dare to be the first. Moreover, managers must not be afraid of failure, or give up in the face of criticism. One of the reasons why Huawei has become a well-known international enterprise is the spirit of innovation that the enterprise group led by Ren Zhengfei possesses. Huawei risked everything by carrying out independent R&D on switches through corporate lending with a high interest rate, and this is rare even in the history of the world. Ren Zhengfei's adventurous and incredible action stimulates the morale of the enterprise's staff. After numerous failures, bearing the loss of from 60 million to 100 million yuan, Ren still firmly continued to invest capital and manpower, and finally they succeeded. Lots of strategic adjustments, such as a market strategy for the transition from rural to urban areas, entering overseas markets, and adjusting the development strategy are inseparable from Ren's and others' spirit of innovation and entrepreneurship.

3. The trajectory analysis of Huawei's independent innovation strategy

3.1 Huawei's R&D strategy adjustment: From technology-driven to market-driven

With the development of Huawei, its R&D strategy has made a transition from "technology-driven" to "technology and market-driven." After failing the bid of China Unicom's CDMA project in 2002,

Huawei realized that pure technological R&D was not beneficial for the company's development when ignoring the market.

Thus, in order to avoid the situation that the researchers only pursue innovative and advanced technology while lacking market sensitivity, Huawei made a hard and fast rule that 5% of its researchers should turn to working on the market each year; meanwhile, some marketers should turn to doing some research work. Huawei not only aims at developing top technology and building a first-class R&D team, but also insists on developing top technology with marketability and on training engineering businessmen, not scientists. Huawei believes that technology is created in order to be sold, and only technology that is sold is valuable. Huawei's successful innovative path is adapting to the market, and not being limited to pure technology and innovation; encouraging innovation, and not innovating blindly.

3.2 Huawei's R&D trajectory adjustment: From a tracking developer to a leading developer

During the early period of Huawei's development, its R&D was mainly tracking development, which means learning from others' mature technology to save product cost and improve competitiveness. However, with the development in recent years, Huawei has reduced the gap with the top international technology. Moreover, it has also gradually become a direct competitor of some multinational corporations. Because of that, Huawei began to suffer blockades and pressure from some competitors. At that time, the tracking development was no longer feasible. Huawei had no choice but to adjust its method to leading development, so as to gain advantages in international competition.

In order to break the the European and American multinational corporations' monopoly on high value-added and high-tech products, Huawei insists on a high starting point in technology research, and always aims at cutting-edge, advanced, and marketable products. It tries to develop at the same speed as some multinational corporations. Nowadays, the leading quality of Huawei's development is acknowledged to all. By continuous R&D and accumulation, Huawei has held a leading worldwide position in the fields of ultra-long-haul DWDM, MSTP, NGN, integrated-service access, IP telecom networks, IP DSLAM, intelligent networks, signaling networks, etc. Besides, it is also assuming one of the world's leading roles in the fields of the intelligent optical network—ASON, switches, WCDMA, CDMA, 3G terminals and so on. From tracking development to leading development, followed by participating in formulating international standards, Huawei's independent innovation has gone further and wider.

3.3 Huawei's market strategy adjustment: From "surrounding the cities from the countryside" to internationalization

In Huawei's early years, there were high barriers to entering the market of large and medium-sized exchange equipment in cities, but the market in rural areas was larger. The products of multinational companies' were high-grade and expensive, and were centered in big cities, less in the countryside and villages. Because of poor traffic conditions and low profit on the rural market, foreign manufacturers were against expanding their market there, but it was an opportunity for domestic telecommunications manufacturers. Following the path of "surrounding the cities from the countryside," Huawei gave its priority to rural markets. After maintaining the company's operation and development, it gradually carried out its marketing activities on urban markets. While Huawei obtained a large market share of the domestic market, it turned to expanding its international market.

3.4 Huawei's strategy of transferring its R&D center: From local development to international strategy

There is a particular characteristic of the communications industry: The one who masters the core technology has a strategic advantage in market competition. Generally, a company considers expanding its market through its products, while Huawei drives the market with its R&D. At the early stage of its development, Huawei focused on building domestic research institutes. Along with the development, Huawei established many institutes in the international market, organizations such as the US Silicon Valley Research Institute, the Dallas Institute, the Swedish Institute, the Indian Institute, the Russian Institute, and international organizations, such as Huawei Technologies Headquater in Shenzhen, etc. Developing the layout of its globalization by localizing the R&D of technology is Huawei's strategy, which is also the nature of the industry. A technical design which cannot adapt to the aesthetics, the way of thinking, and the working habits of the local people would not win over the hearts of customers. The products and technologies without complete localization and integration would be more difficult to gain the customers' love and loyalty. Based on the deep understanding of the nature of the industry, Huawei shaped a globalization strategy led by R&D localization.

The indirect strategy of "borrowing ships to sail" is Huawei's preempting plan. Customers in developed countries are unfamiliar with Huawei, but few people are unfamiliar with Nokia, Ericsson, Siemens, GE, Microsoft, IBM, SONY, Alcatel, 3COM, etc. Huawei and these tycoons have shaped strategic alliances, and to a certain extent, customers get to know something about the brand of Huawei through these tycoons. Moreover, Huawei recognized the value of indirect strategy, so it implemented extensive cooperation with multinational companies. With its own products and technologies, a way of equity participation or main holding and with the help of well-known brands and channels, Huawei has expanded into the global technology territory based on its own brand of technology.

Enterprise Case 2

CIMC—Independent innovation forging a world-class enterprise

1. Rapid development of CIMC

China International Marine Containers (Group) Ltd. (CIMC in short) was founded in 1980. It is one of the earliest professional Chinese container manufacturers and Sino-foreign joint ventures. At present, the main business of CIMC includes the designing, manufacturing, selling, and maintaining of various kinds of containers, special vehicles, and semi-trailer trucks as well as other transportation equipment. In 2015, 60,000 excellent staff delivered sales revenue of 58.69 billion yuan and net profits of 1.97 billion yuan. Its total assets reached 106.76 billion yuan, and net assets 35.49 billion yuan. CIMC has more than 300 member enterprises and 3 listed companies in Asia, North America, Europe, Australia, and others. CIMC is dedicated to supplying high-quality and reliable equipment and services of modern communications and transportation. After a 36-

year period of development, CIMC has become an outstanding global container manufacturer that gradually rose from a traditional product manufacturer (by 2016). There is no doubt that independent innovation is the key to its advantages in global competition.

1.1 Leading the world's container manufacturing industry

Chinese container manufacturing started in the early 1980s, more than 20 years later than the emergence of the world container manufacturing industry. To create a new industry, it requires a process of introduction, implementation, and improvement. From its production in 1982 until 2016, CIMC has gone from a small enterprise on the verge of bankruptcy to a leading world-class container manufacturing group. CIMC's developmental and value-added trajectory have gone through the process from surviving by cost advantage to developing by independent innovation.

CIMC's initial stage lasted from 1980 to 1992. As one of the earliest joint ventures after the implementation of reform and opening up, the purpose of CIMC's founding was to introduce international container manufacturing technologies and management experience. However, because the foreign side lacked understanding of Chinese national conditions, which led to cultural conflicts within the enterprise, CIMC suffered continual losses. In 1987, China Ocean Shipping (Group) Company (COSCO in short) joined CIMC, which is considered a turning point in CIMC's development. With the recovery of the international container industry, COSCO continued its reorganization within CIMC. Thus, the company's capital increased to $4 million, and COSCO and China Merchants accounted for 45% of total shares respectively, while the East Asiatic Company (EAC) accounted for 10% of total shares. As COSCO's supporting base, CIMC made its way to the market and won the first batch of container orders after its reorganization.

The stage of cost leadership strategy occurred from 1993 to 2001.

With the continuous strengthening of global economic integration and the increasing global trade volume since the 1990s, the demand for container update increased steadily, and that for containers maintained a growth rate of 8% to 9% every year. CIMC realized that China had a huge container market, and successively acquired and merged more than 10 container production enterprises in coastal areas across the country. Moreover, CIMC set up a production layout system covering the major coastal ports in the country earlier than its competitors. Thus, it shaped its production base according to the market and scale advantage. At the same time, CIMC unified its bulk material purchasing, order sales, marketing, and financial management through the enterprise group management operation mode. By 2001, CIMC had produced series of mainstream special containers. The sales volume for special containers reached 26,317 TEUs in 2001, a 78.86% increase over 2000. The sales volume in 2011 accounted for 5.81% in the world; compared with 2000, the proportion had increased sharply.

Since 2002, CIMC has been in an independent innovation and global operation stage. In 2003, the sales volume of CIMC containers was more than 1 million TEUs for the first time and the sales revenue was over 10 billion yuan for the first time; CIMC created the first two annual operation indexes in the global container manufacturing industry. Currently, in all turnover containers in the global cargo system, one in every four containers are made by CIMC, and one of every two newly-made containers are CIMC products.

“Resource globalization, service internationalization” is CIMC’s management principle. CIMC has set up and perfected an international customer service system, a joint venture production system, a technical service system, a procurement system, and a financing system. In the process of world economic integration and global industrial chain integration, depending on a steady and bold management style, clear cognition about the changing era, and a firm grasp of historical opportunities, CIMC enhances its international competitiveness by integrating industry resources with a global perspective, taking full advantage of China's comparative advantages, and strengthening its cost control and management innovation to enhance its international competitiveness.

1.2 Successful entrance into the semi-trailer and special vehicles manufacturing industry

The container industry and the semi-trailer business have similar industrial patterns; the latter is the extension of the former's core competitiveness; both have strong strategic synergy. CIMC's advantages and strengths regarding technology and management platforms, sales networks, and the brand resources of the container industry gradually extend to the semi-trailer and special vehicles manufacturing industry. Along with the networking of China's highway system, the growth of customer demand for a rapid freight system and the reform of the organizational mode of logistics service as well as national policy support for the special-purpose vehicle and semi-trailer industry, the trend in China has been towards a large-scale commercial application of special-purpose vehicles and semi-trailers. In this context, CIMC started to produce semi-trailers and special-purpose vehicles.

At present, semi-trailers are considerably popular in Europe, America, Australia, and other developed countries and regions, and are the main force of inland freight transport. In general, medium and long-distance highway transportation can be undertaken by semi-trailers and other large transport vehicles. In contrast, since China's high-grade highways are paid to use, the semi-trailer fails to fully tap its market potential. The goal of CIMC's semi-trailer business is through domestic cooperation and overseas introduction in order to provide the global market with first-class road transport equipment, and in turn, to introduce an advanced worldwide road transportation mode to China. At the same time, CIMC is engaged in promoting the modernization of Chinese transportation equipment.

The special-purpose vehicle is a kind of highly efficient, safe, reliable, large mass-bearing road transport machine, which is widely used in Europe, the US, and other developed countries and regions and whose market scale is 3–4 times that of the container. CIMC expands its semi-trailer and special-purpose vehicle business through M&A and purchasing. At home, CIMC has finished the first phase of strategic acquisition, stationing, and preparatory engineering construction work of CIMC Tonghua Company (Yangzhou), CIMC-Koegel Company (Jinan), and CIMC Juki eastern factory (Shenzhen). Abroad, CIMC has carried out extensive business cooperation and has formed technical alliances with some well-known enterprises from the US, Japan, Germany, UK, and other countries. For example, CIMC bought assets of the fifth largest semi-trailer manufacturer, HPA MONON of the US, in 2003. This action provides multimodal transport, road transport, railway transport, and the third party logistics mode with a personalized and modern transportation equipment solution. At the same time, CIMC can learn advanced foreign management and gain marketing experience.

Through re-innovation of introduction, adoption, and implementation, and integration innovation to original innovation, CIMC takes the world's leading position in transport equipment on the three levels of products, patents, and standards; through technological innovation development strategy, with the protection of IPR as its core, CIMC has shaped the patent management system of declaration, implementation, rights protection, and defense, and built an intellectual property system with patents, trademarks, and trade secrets as its main contents. It is the constant innovation achievements that lead CIMC to be a leading enterprise in the container industry.

2. The characteristics of CIMC's independent innovation mode

2.1 CIMC's independent innovation mode: M&A and independent R&D

M&A is a shortcut to obtain technologies; independent innovation based on M&A is the core of the sustainable development of enterprises. By means of M&A, CIMC creates the myth of becoming an industrial leader. CIMC acquired 14 enterprises and trusteed a container manufacturing enterprise; thus, CIMC not only realized the strategic layout of the container industry, but also defeated a lot of competitors. CIMC entered the special-purpose vehicle industry in 2002; for three years, it had acquired four domestic enterprises and one American enterprise, and quickly became the largest special-purpose vehicle manufacturing enterprise in China. Through continuous M&A, especially for technical resources, CIMC rapidly achieved scale merit and laid the foundation for becoming an industrial leader.

In the process of M&A, choosing to use the most valuable international resources becomes the model of CIMC internationalization, and it also provides the trajectory to successfully achieve Chinese enterprise internationalization. The M&A of CIMC is not simple scale M&A; through the integration of foreign brands, the technicians, and the enterprise, bypassing market and technical barriers, CIMC relied on high quality international resources to integrate and gather global container resources and in this way, entered the high-end market rapidly and defeated its competitors to become one of the worldwide enterprises with advanced technology.

2.2 CIMC's independent innovation trajectory: From technology purchase to imitative innovation and standards formulation

CIMC's great performance are attributed to a clear trajectory of technological innovation; every one of its major independent innovations has gone through three stages from imitation to learning and independent innovation. In 1995, CIMC decided to produce standard coolers. 95% of the coolers in the market were aluminum. The core technology of stainless steel coolers was controlled by the Germans, and that of aluminum coolers was controlled by the Japanese. CIMC believed that the steel container was the future direction, so it bought German patent technology and began to produce steel coolers in Shanghai. However, CIMC's purpose was to develop more advanced technology. Based on the German technology, CIMC carried out technology upgrading and independent R&D, and applied for 11 patents, which replaced the imported technology. With the improvement of CIMC's technological innovation ability, CIMC gradually participated in formulating international technology standards. At the TC104 meeting, CIMC plays a dominant role.

While gradually obtaining key technology IPR in various fields, CIMC protected them in accordance with international rules, and carried out a number of actions for rights protection; for example, cases for vehicles and coolers achieved some progress and had a good effect on society; this series of actions have played an important role in guiding the industry to respect IPR and in promoting the healthy industrial development.

2.3 Special ownership structure, the basis of CIMC's independent innovation

The ownership ratio determines the decision-making right and the initiative of the controlling shareholders in the process of decision-making. At present, the Chinese side of some domestic backbone enterprises is losing their control in sino-foreign joint ventures and cooperation; some are even strictly limited by the foreign holdings. However, in line with the following three principles, CIMC has gained advantages in the ownership structure.

First, the principle of check-and-balance of the ownership. After its reorganization, there are three shareholders: China Merchants with a 45% share, COSCO with a 45% share, and EAC, a Danish company, with a 10% share. EAC's share shapes a check-and-balance relationship between the two major shareholders to prevent the shareholders from damaging the interest of the entire enterprise or minority shareholders due to the pursuit of their own interests; this ensures a sustainable and healthy development of CIMC.

Second, the principle of "common language." CIMC's products were integrated into the international market at the very start, and all the shareholders are familiar with international market operation and have few significant divergences in decision-making, which reduces the decision-making cost and improves the decision-making efficiency.

Third, the principle of "resource integration." CIMC did not operate smoothly when it was founded. At the beginning, CIMC showed a deficit. The turning point was the participation of COSCO, which effectively achieved resource integration and advantage complementarity. COSCO not only increased the capital investment and expanded the equity, but more importantly, as the world's fifth biggest international container shipping company, COSCO became a user of CIMC containers, which expanded the enterprise's market.

The three principles above ensure the decision-making power of CIMC for independent innovation, and make it impossible for CIMC's independent innovation to be submerged in the process of M&A, joint venture, and cooperation.

3. Summary

The strategies of CIMC are formulated from a global perspective to establish and develop the following three parallel levels: The first level is the current core container business; the second level is the modern road transport equipment business; the third level is to participate in the vital businesses of the modern transportation equipment and service industry in a broader sense. Through independent innovation, CIMC moves far ahead on the road towards container industrialization.

Enterprise Case 3

BYD—From a battery king to an auto manufacturer

1. The miracle from 2 million yuan to 80 billion yuan

BYD Co., Ltd. (BYD in short) is a Chinese private enterprise, which was listed in Hong Kong, and has two major industries: IT components manufacturing and auto manufacturing. The main products of the former include lithium-ion batteries, nickel metal hydride (NI-MH) rechargeable batteries, and LCD screens; the automobile products cover all kinds of high-end and low-end series of cars from 800 cc to 2,400 cc as well as electric vehicles and electric bikes. Since the company's foundation in 1995, its annual sales have grown at a speed of more than 70%. BYD's annual sales revenue were about 80 billion yuan in 2015, and the net profit reached 2.89 billion yuan. The market share of lithium batteries, nickel batteries, and buttons for mobile phones has reached the top one worldwide. One out of every four mobile phones use a battery produced by BYD. BYD has become a world-class mobile phone components manufacturer and solution provider, and is developing rapidly in the auto industry as well. In its over 20 years of development, BYD has won out over SONY, Toshiba, and other world-class competitors, and grown from a pure manual and mill-type domestic enterprise with an initial capital of only 2 million yuan to a listed enterprise with sales revenue of 80 billion yuan. Quality, cost and efficiency are the key to the success of BYD. As for quality, the foreign leading enterprises have certain advantages; while concerning cost control, domestic enterprises are in a more favorable position. BYD tried to gain a market share on the basis of ensuring quality and reducing cost. BYD's advantages regarding quality, cost, and efficiency are established on the basis of innovation; and independent innovation is the focal point of its rapid development. It is the independent innovation that has provided BYD with core technologies and products with independent IPR and has allowed it to become a competitive multinational company born out of a humble private enterprise.

1.1 Seizing business opportunities by starting from a low-end product—the battery, BYD became a battery king within 8 years

In a span of eight years, BYD became a leading battery producer that started out from a humble private enterprise. In 1995, BYD was founded with a registered capital of only 2.5 million yuan and 20 staff. In 1997, BYD began to produce lithium-ion batteries. Through the optimizing of the negative electrode additive and through independent innovation, BYD improved the performance of low-current charging and the battery's life cycle, which boosted the sales. Thus, BYD grew to a medium-sized enterprise with annual sales of nearly 100 million yuan. In 1999, BYD's nickel-cadmium (NI-CD) battery production reached 150 million, 250 million in 2001, and 400 million in 2003, exceeding Sanyo, and becoming the battery king. In a short span of eight years, BYD had become the battery supplier for well-known mobile phone brands, such as Nokia and Motorola. BYD's success comes from its tireless innovation and its innovation strategies in different periods.

In the early 1990s, Japanese battery manufacturers, such as Sanyo, Toshiba, and Panasonic, occupied nearly 90% of the global market share. BYD used secondary rechargeable batteries as a breakthrough to enter the battery market since they had a low entry barrier. Insisting on technological innovation, BYD used the low-cost strategy to produce batteries manually. Japanese companies used an automatic production line to produce batteries at a high cost, while BYD adopted the manual work mode and divided a production line into several artificial procedures. When launching new products, the original production line just needed some partial adjustments in key links, combined with corresponding technical training for staff, so the cost of production was reduced greatly. BYD survived the 1998 financial crisis with the low-cost strategy, and occupied 40% of the global market share of NI-CD batteries. Subsequently, BYD developed rapidly, and became a real battery king in 2003.

1.2 Entering the auto industry, BYD realizes its auto dream

In 2003, BYD spent 269 million yuan on purchasing 77% of the shares of Xi'an Qinchuan Automobile Co., Ltd., which was the only authorized large modern enterprise engaged in car production in northwestern China. So far, BYD has realized an interactive platform for the battery, which is the main business, and the electric vehicle, and become the second largest private domestic enterprise after Geely.

Unlike Bird and Aux which rushed in and out of the auto industry, when BYD entered the auto industry, it conducted internal adjustments and resource integration. At the same time, BYD carried out technical R&D and spare parts supporting, and expanded the network in a planned way. In 2003, BYD bought the mould company of BAIC Group, and established the Beijing BYD Mould Co., Ltd., so it had the ability to reduce the cost of mould production and to improve the quality of car body. In April 2005, Shanghai BYD Automobile R&D Center was established; there were more than 20 project teams that were engaged in the research and exploration of the BYD car body series, car electronics, security devices, electric cars, and other aspects. BYD also declared more than 100 patents successfully in half a year. Subsequently, BYD established Shanghai Auto Industrial Park, and a complete framework for an automobile R&D system, the vehicle inspection center, a collision line, and all kinds of environmental laboratories. After finishing the production, R&D, and sales layout, BYD launched the first F3 car, which won market recognition due to its good cost performance; and BYD successfully completed the dual task of expanding its market and building its brand. In the first half of 2006, BYD sold 32,500 F3 cars, with an annual increase of 614%. Subsequently, BYD launched the F3 Honorable and F8 hardtop roadster. In 2006, its total sales volume were up to 63,153, a yearly increase of 472%, and thus it became the fastest-growing independent auto enterprise, getting the "2006 China Top 10 Marketing Events" prize. In 2007, BYD signed an export agreement with Portugal and other countries, and its journey of auto internationalization began. In 2008, BYD won "The 2008 No.1 Brand of Chinese Market Buyers' Satisfaction" award, because of its brand leadership in the industry and its good brand awareness. In 2009, BYD successfully bought the full ownership of Hunan Midea Coach Manufacturing Co., Ltd. In 2010, BYD ranked No.8 in the world's top 50 most innovative enterprises list, being the highest-ranking Chinese enterprise; moreover, it ranked No.68 in the top 500 Chinese brands. BYD, the only Chinese brand listed, ranked No. 466 in the top 500 global brands in 2011.

The key to BYD's successful entrance into the auto industry is that BYD combined the battery and the auto industry to develop electric vehicles. The bottleneck of electric vehicles' development is the car battery. The traveling distance with one charge of it, its life, and its recycling are the problems of electric vehicle battery. As the battery king, BYD has a good foundation in terms of batteries. If BYD insists on technological innovation and has a breakthrough in battery research, it will be able to gain the initiative in the competition in the electric car market. At present, the key to BYD's success of electric vehicles is its iron battery that costs less than lithium battery, but has almost the same capacitance as the lithium battery, and will not explode in the process of combustion and collision, and is much safer. In addition, iron battery can be recycled 2,000 times, guaranteeing the electric vehicle a driving distance of 600,000 km, and can be used for ten years. It has become the core technology of BYD, which has applied for more than 630 patents at home and abroad for its electric vehicles equipped with batteries. BYD has taken a crucial step in the field of electric vehicles.

2. The experience of BYD's fast growth—Independent innovation

The key to BYD's success can be attributed to the following factors.

2.1 Persevere independent innovation

BYD's independent innovation is embodied in various aspects. BYD is not only actively involved in the fields which have high innovation requirements, such as new product development, but the company also carries out innovation in the production of raw materials and in production line modification. While continuing to improve the quality of its products, BYD tries to reduce costs as much as possible and to improve its competitiveness.

At the beginning of its establishment, due to the lack of work shops, equipments, and raw materials, BYD considered technology as its lifeline, and thus it made huge investment in the most advanced equipments and talents, and set up a central institute, a design studio, and a technology department which would be responsible for the R&D of the production equipment and process of secondary rechargeable batteries. BYD's innovation is not merely imitation, but a trajectory suitable to its own development.

Regarding technology R&D, one project would be simultaneously developed by different technology groups. The team which first proposed an applicable idea could be the first to begin its production. This innovation system allowed BYD to gain time and be able to catch up with the Japanese battery giants.

Regarding production line innovation, BYD did not blindly imitate the automatic production lines of other giants, such as Sanyo, instead, it used mainly a manual production line that was more suitable to their own financial conditions. BYD's nickel battery production lines and lithium battery production lines were all developed independently according to performance requirements and low cost principle, which had saved the company a lot of money.

Concerning the innovation of raw materials, BYD was directly involved in the development chain of suppliers, and, together with them, worked out plans to reduce the cost. For example, BYD cooperated with suppliers from Shenzhen to improve the quality of cobalt, the raw material for NI-CD batteries and to reduce its cost. In this way, BYD was able to save tens of millions of yuan every year. The modification of nickel materials had also permitted BYD to reduce the cost of nickel from 5–6 million yuan to hundreds of thousands of yuan.

The reason why BYD has the capability for technological innovation and constant innovation achievements is that the company has established a mature multi-level technological innovation system. BYD has established a tertiary R&D system, including a central institute, a research department, and a technology department. The central institute mainly carries out strategic preparatory work and develops high-level patent achievements, and then developers enter the research department for further R&D with the developed achievements.

After the achievements become mature through previous development, other patent achievements will be developed to produce corresponding products. The establishment of a perfect multi-level technological innovation system makes BYD's technological innovation truly realize the strategy of "one generation of pre-research, one generation of reserve, and one generation of application."

2.2 Persistent innovation of management and market development strategy

While insisting on independent innovation, BYD pays attention to the innovation of its management and market development strategy, so as to better play the role of technological innovation. On the one hand, BYD builds a strategic alliance relationship as much as possible with its suppliers. On the other hand, BYD keeps on developing its market.

BYD's customer-oriented marketing strategy can be divided into three levels: The first level is the alternative market, which mainly includes small manufacturers, whose technical and quality requirements for products are not so high, but which do hope for prices that are as low as possible; the second level is the domestic manufacturers, which require higher quality and lower prices; the third level is the high-end market, such as DBTEL, Lucent, Motorola, and other international giants, which have very high requirements regarding the suppliers' quality, technology, and R&D capabilities. In the process of market development for every new product, BYD does not aim at immediately entering the third-level market; instead, it gradually develops its market from the first level to the third level. Through the experience of the first two levels, BYD accumulates enough experience and strength to enter the third-level. This progressive market innovation permits BYD to enter the advanced ranks of the international market step by step.

2.3 Reasonable combination of "labor-intensive" and "technology-intensive"

BYD's independent innovation is not only shown in its

technological innovation, but also in its management mode. Its own "labor-intensive plus technology-intensive" mode explores the road towards BYD's rapid development. In its development, BYD did not follow footsteps of many domestic enterprises, which spent a lot of money on importing automatic production lines from abroad; instead, according to its own characteristics, in combination with Chinese national conditions, BYD reasonably matched up high-quality human resources with capital-intensive equipment, and formed a unique developmental mode. This kind of developmental mode can quickly adapt to rapid technological changes. Setting a production line independently, freely increasing the scale of production, and adjusting the product specifications in order to adapt to the needs of customers quickly are its obvious advantages, especially during the financial crisis. After the financial crisis, the global prices for battery products collapsed, and many Japanese manufacturers were difficult to survive, but BYD was bucking up; its growth rate was as high as 90%. BYD received the orders that previously went to Japanese manufacturers, and expanded its market share to gradually become a strong performer in the battery industry.

3. Learning from BYD's independent innovation

BYD grows from a private enterprise set up in the mid-1990s to the world's largest rechargeable battery manufacturer in a span of more than ten years. BYD's success makes us realize that:

(1) Private enterprises are an important force which can be relied on to improve the innovative ability of China. Private enterprises have unique competitive advantages, strong flexibility, good innovative culture, and strong sense of innovation in their development. Therefore, it is necessary to let private technology enterprises play an important role in improving China's capability for independent innovation and national competitiveness.

(2) The developmental mode of the enterprises in developed countries cannot simply be imitated. BYD's experience suggests that labor-intensive and technology-intensive mode have complementary advantages and their combination is able to bring China's comparative advantages in human resources into full play. Blindly imitating the highly-automated production mode of developed countries is not advisable.

(3) Owning independent IPR is a powerful weapon in winning out over the competitors in the international market. The competition between BYD and Sanyo regarding IPR fully illustrates that only by owning independent IPR can an enterprise remain in an impregnable position in a market with fierce competition.

Enterprise Case 4

Tencent—Winning by "Q"

Since its birth in 1988, "QQ" has attracted hundreds of millions of users, and has become an Internet product with widespread brand popularity and huge value potential. What kind of independent innovation strategy supports the rapid rise of the Internet giant, Tencent? Tencent is the earliest Chinese Internet instant messaging software developer and a provider of Chinese Internet and mobile value-added service, and has been engaged in the service operation of instant messaging and related value-added business. At present, Tencent is one of the largest Chinese Internet comprehensive service providers and one of the Chinese Internet enterprises which serves the most users. Its development involves Internet value-added service, mobile and telecom value-added service, and Internet advertising service. By the Chinese leading network platforms, such as the instant communication software (QQ), Tencent website (QQ.com), Tencent games, QQ space, wireless portal,

SOSO, PAIPAI, Tenpay, etc., Tencent tries to meet the Internet users' need for communication, information, entertainment, e-commerce, etc. In the first half of the year 2016, QQ's monthly active accounts reached 899 million, and the peak of online accounts, at the same time was 145.4 million.

It is Tencent's mission to use advanced Internet technology to improve the quality of human life. The development of Tencent's QQ has profoundly impacted and changed the method of communication and the living habits of hundreds of millions of people; it provides the users with a huge convenient communication platform. Moreover, it provides all kinds of life functions, social service functions, and business application functions for the life of human beings; at an unprecedented speed, QQ is changing people's life-style and creating a broader Internet application prospect.

1. The history of QQ's development

1.1 The beginning

In November 1998, Ma Huateng and Zhang Zhidong created Shenzhen Tencent Computer System Co., Ltd. with capital of 500,000 yuan, and launched a wireless Internet paging solution (TIPS). Its main target was to connect Chinese paging users with Internet users, to provide more value-added services and provide consumers with an Internet "wireless gateway." In 1999, Tencent officially launched its independently developed instant communication service (Tencent Instant Messenger), and released the first test version of the QQ99b0210 software, or Tencent QQ for short. With the growth of its service users, Tencent gradually developed a wide range of Internet business, including e-commerce, integrated information services, and advertisement. In the initial development period, Tencent concentrated its target of enterprise development on developing and expanding a unified, highly speedy, rich Internet application platform and on making an effort to expand online advertising, mobile communication, ISP, and the IP telephone cooperation business, enterprise business instant messaging, and online customer support, as well as online calling business.

1.2 The developmental period of "Alliance cooperation plus technological innovation"

In the period from 2000 to 2005, Tencent expanded its areas of business and obtained a rapid increase in the number of its users through cooperation with China Unicom and CMCC. In May 2000, Tencent cooperated with Shenzhen Unicom, and launched the latest mobile communication service—mobile QQ, which realized the exchange of information between mobile phones and Tencent QQ users via the short message server. This greatly facilitated online contacts for mobile phone users; once the phone users of Shenzhen Unicom openeds the function of short message service, they could use their mobile phones to send and receive QQ short messages or to search for QQ friends as well as to send and receive emails, etc. Even if the users are traveling, they can keep in touch with QQ friends at any time; mobile office has truly been achieved. Through cooperation with China Unicom, this technology has helped Tencent and China Unicom win a large number of loyal customers. While receiving favorable reviews from their customers, these two companies achieved a win-win situation. During this period, the number of QQ online users reached 100,000 for the first time.

In August 2000, the "mobile QQ" service was put into trial operation by Shenzhen Company of Guangdong Mobile, and Tencent QQ achieved a major breakthrough. The Tencent QQ2000 version was released in November 2000, which had a greater improvement than the old version with richer functions. Concerning customer needs, QQ2000 tended to have a more humanized design, and many more new functions, such as skin changing, information providing, Tencent Browser. The strength of the enterprise increased together with its development. In January 2001, the world-famous Internet special research firm NetValue announced the ranking of Internet websites and entities of five Asian countries and regions. Tencent ranked No. 6 after Sina, 163, Sohu, Capital Online, and Fm365 (.com).

Tencent QQ has been launching new business continuously. After QQ became the first domestic Internet service with 100,000 online users in 2000; in February 2001, the number reached 1 million. Within eight months, QQ had reached a new peak again. In March 2002, the number of QQ registered users reached 100 million. In September 2002, in the ranking list of "Monternet SP" published by Zhejiang Mobile, Tencent QQ ranked No. 1 and won the first place in the SP evaluation in the second quarter of Zhejiang Mobile. This once again proved the users' approval to QQ. In addition, Tencent is also committed to cooperating with China Telecom to expand the Internet business together.

In September 2003, Tencent launched its core technology brand Real Time eXchange (RTX), which showed Tencent's strength and determination to enter the enterprise real-time communication market. However, the potential of the market was tremendous and grew rapidly; the realization of a "real-time enterprise" was a long-term goal, and Tencent would not be able to develop it alone. Therefore, Tencent continued to look for alliance cooperation in order to develop a real-time industry value chain with lots of enterprises. Tencent upheld an "open and collaborative" attitude to establish strategic alliances positively with other outstanding IT manufacturers and important technology partners like Yonyou, Kingdee, and Kingsoft. Tencent established a strategic cooperation relationship with the blue giant of information industry—IBM to enter the enterprise real-time communication market hand in hand. In December 2004, Tencent declared that it would take a "horizontal plus vertical" business mode on the basis of profound integration; the original instant communication tools (QQ, TM, RTX) were the vertical core, while the QQ.com web portal was the horizontal center. With the superposition of two dimensions, Tencent's business would cover wireless value-added service, interactive entertainment service, Internet content service, and B2C, C2C e-commerce, as well as auctions. This kind of developmental mode can not only serve more users, but also carry more cutting-edge technology. Based on this, it could expand a broader network value-added service to meet and satisfy the users' increasing demand for a network.

In 2004, the number of Tencent QQ registered users broke the 300 million threshold, and the enterprise had developed by leaps and bounds with nearly 30 software products with copyright. Tencent was listed in "Shenzhen Top 50 Private Enterprises in 2003," and ranked No.4 in list of "Top 100 Shenzhen Software Enterprises in 2003" and No.17 in the list of "Top 500 High-tech Enterprises of the Asia-Pacific Region in 2004." In 2005, Tencent QQ's peak concurrent users surpassed 10 million for the first time, 100 times that in 2000. Tencent ranked No. 25 in the list of "The 100 Largest Enterprises in the Chinese Software Industry in 2005." In a word, Tencent expanded its business and attracted users through constant cooperation and innovation.

1.3 Tencent's rapid growth: Since 2006

In 2006, Tencent was identified as a "Shenzhen enterprise with advantages in intellectual property," a key software enterprise for the national planning layout of 2005, and an independent innovative pilot enterprise in Guangdong Province. In 2007, with the tenet of "independent innovation and focusing on practicality," Tencent was committed to developing businesses by extensive R&D strength and operational capabilities as well as excellent technical innovation ability; thus, it won the "Private Enterprises' Innovation Award in 2006." The Chinese brand "腾讯" and the English brand "Tencent" got the title of "Famous Trademark of Guangdong Province" due to its absolute advantages of high quality, good reputation, and large market share. In the same year, Tencent QQ's peak concurrent users reached 30 million. In 2008, Tenpay won "The Most Trustworthy Payment Platform by B2C Merchants Award" for its secure and convenient payment service and leading market share. Tencent's "penguin" platform strategy mode developed a harmonious road for business and environment, and realized the healthy development of market and industry, and thus it was awarded "The Best Business Mode in 2007" and the title of the third batch of "National Cultural Industry Demonstration Base" by the Ministry of Culture, and became the only winner among Internet comprehensive service enterprises. In 2009, Tencent's total number of authorized patents surpassed 400, and it became one of the global Internet companies with the largest number of patents; the "Tencent QQ" trademark was certified as a well-known trademark by the State Administration for Industry & Commerce; Tencent formally obtained a national high-tech enterprise certificate. At the same time, Tencent QQ's peak concurrent users surpassed a new threshold again, 50 million. The users logging into QQ space reached 200 million, and thus, Tencent kept on being one of the world's largest Internet social network communities. In 2010, Tencent QQ's peak concurrent users reached 100 million.

Innovation is the core strategy for Tencent's survival and development. After over a decade of experimentation and practice, Tencent has gradually finished the entire developmental process from learning innovation, integrated innovation, to strategic innovation. Tencent was devoted to broadening its business through the "alliance cooperation plus technological innovation" developmental mode in its early and middle development. While insisting on the "alliance cooperation plus technological innovation" mode in its rapid development, Tencent gradually combined technological innovation with brand and management innovation in search of a developmental mode suitable to its own development.

2. The innovation path of QQ

2.1 "Improving technology plus venture investment" to make Tencent rise

Starting a business often depends on an entrepreneur's sensitive grasp of the new demand and knowledge. It is a kind of innovative activity where the previous technologies are surpassed due to the entrepreneur's improvement, and the new technologies will be applied in a different environment. Tencent QQ's innovation involves catching up with and improving the previous technology.

Tencent QQ developed from the ICQ, a software invented by the Israeli company Mirabilis in 1996 for people's rapid communication on the Internet. As one of the ICQ users, Ma Huateng found that the English interface and functioning of ICQ was not suitable for Chinese, so he developed the product OICQ (similar to ICQ) that was suitable to the habits of the Chinese user. With the rapid increase of OICQ users, Ma Huateng and Zhang Zhidong realized the market prospects for this technology, and therefore they founded Tencent, and insisted on continuous innovation in order to meet market and customer demands. They have developed QQ, RTX, IM, the web portal QQ.com, and other products.

If it is the technology that promoted the establishment and development of Tencent, capital is the indispensable key to its development. Faced with the failure of being merged and difficult domestic financing, Tencent got financing through foreign venture capital, which consequently promoted its development.

Concerning the innovation mode, Tencent's innovation process contains technology push and market pull. At the very beginning, Tencent QQ opened instant messaging technology for netizens, and promoted market demand to bring consumers convenience. The increasing number of Internet users and more demand for new technology encouraged Tencent to develop RTX and IM. Thus, Tencent further improved its business scope which in turn promoted its development.

2.2 Independent IPR that consolidate the status of the independent brand "Tencent"

Independent IPR is the key to the growth and survival of enterprises, especially for high-tech ones. At the beginning of its establishment, the maintenance of IPR was still insufficient. The domain name "oicq.com" of OICQ was accused of infringement by international IM software who directly required international domain name registrars' compulsive transfer. Therefore, Tencent changed its brand name from OICQ to Tencent QQ. Armed with the consciousness of its trademark and patent registration, Tencent changed a batch of cartoon images. The previous symbol of the little penguin similar to that of ICQ was replaced by a little penguin in a red scarf to avoid the dispute of IPR. Later, Tencent started to strengthen the position of its brand and to advertise the small cartoon image of penguin as the QQ logo. In the trademark fight, Tencent gradually realized the importance of IPR, and positioned itself as an instant messaging and value-added services operating company, which could be visited through a variety of communication platforms (such as PC, mobile phone, fixed-line, household appliance, and various kinds of communication terminals) based on the Internet. Along with the growth of the enterprise, Tencent actively implemented and promoted its overall strategy of IPR, including patents, trademarks, and copyrights. Meanwhile, Tencent actively participated in various association activities in the field of IPR, so as to improve the competitiveness of the enterprise and strengthen its status as a leading Internet enterprise.

2.3 Developing products that meet the users' needs to promote Tencent's growth

As the number of users surged, the cost of Tencent also increased rapidly, which gradually consumed all of Tencent's capital. Failing to control the user registration, Tencent sought a new way to make a profit suitable to its own characteristics. Through cooperation with the traditional communications industry, Tencent divided QQ's business into QQ download, QQ community, QQ member services, mobile QQ, enterprise QQ, QQ game, and advertising services in order to promote Tencent's growth.

3. The characteristics of Tencent's independent innovation

3.1 A good innovation environment to provide Tencent with an innovation platform

Enterprise independent innovation needs a good social environment; the government's behavior and attitude determines the development of innovative enterprises. The establishment and growth of Tencent cannot be separated from Shenzhen's social environment. Shenzhen municipal Party committee and municipal government did almost everything possible to promote independent innovation; they formulated a series of policies and regulations to encourage innovation, provided innovative companies with capital, land, and other preferential policy support, and actively introduced talented people, providing a good environment for them; moreover, they established and improved an investing and financing system to promote the development of independent innovative enterprises, attached great importance to the protection of IPR, and introduced well-known domestic and overseas universities to establish virtual university parks. Shenzhen has established a regional independent innovation system which takes the market as the leader, industrialization as the purpose, enterprises as the subject, and government–industry–university–research as a close alliance. Tencent is growing in the good innovation environment of Shenzhen.

3.2 Team leaders' entrepreneurship, the backbone for Tencent's growth

Entrepreneurs with a spirit of innovation are a key factor in promoting enterprise growth. Throughout the development of large domestic and foreign innovative enterprises, there can be no development without the entrepreneurs' spirit of innovation. The entrepreneurs with a spirit of innovation can make a choice that allows the enterprise that is at a critical moment of its development to change its strategy so as to point in the right direction for development; therefore, they are the key to promoting a vigorous development of the enterprise. Ma Huateng from Tencent is a representative figure; he won the fourth "Bauhinia Cup Outstanding Entrepreneur Awards" issued by Hong Kong Polytechnic University, and was included in the list of the "Global Young Leaders." Tencent's founding leadership team won the "Special Contribution Award of Corporate Citizenship" by their outstanding contributions to carrying out social responsibility. It is also the only award given to the Chinese Internet industry. It is because of the emphasis on independent innovation and outstanding entrepreneurs with high strategic vision that Tencent can go farther and farther along the road to development.

3.3 Independent R&D and innovation, the perpetual motion machine for the sustainable development of Tencent

The rapid development of Tencent in the Chinese Internet industry is attributed to its perception of user needs and innovation. With the development of the enterprise, Tencent has formed a set of relatively complete innovation mechanisms for updating and developing itself, and a corporate culture which encourages employees to innovate. The company has not only set up a substantial prize to reward the employees who have carried out all-around innovation, but also held an "innovation competition" at regular intervals within the company to award excellent individuals and teams. An excellent enterprise innovation environment should provide, as does Tencent, the best soil for the inoculation and growth of technological innovation. Outside the enterprise, Tencent actively carries out innovation activities, such as holding innovation competitions, with the topic "innovation creates value," to encourage the public to participate in Tencent innovation. In terms of technology productization, Tencent has established a set of rapid application channels of basic technology. It has formed a set of sound pyramidal R&D structures inside the company, including the research institute, innovation center, and product business department; this three-level R&D system involves different stages. While boosting the R&D of six core technologies, the productization application can also be guaranteed.

3.4 Starting high-end R&D, focusing on core technology, and entering into the core field

At present, research on the Internet in China is focusing on the application field, which directly leads to inadequate control of core basic technology for the Chinese Internet. While facing globalization competition in the Internet field, we are often just passive followers. Therefore, Tencent has established some research institutes to focus on basic practical Internet technology and has been carrying out high-end R&D through independent innovation, and finally enters into the core technology field of the Internet. In addition, the establishment of Tencent research institutes also shortens the time to make the latest technology popular with the users, which permits the terminal users to quickly enjoy further excellent experience from technological innovation.

Setting up research institutes is one of the measures which international high-tech enterprises often take to guarantee their technical innovation ability. Currently, international Internet companies like Microsoft and Google, and domestic enterprises like Lenovo have established their own research institutes. Tencent research institutes are the first established independently by a domestic Internet enterprise. They are in Beijing, Shanghai, and Shenzhen, and their research emphasis focuses on storage technology, data mining, multimedia, Chinese language processing, distributed network, wireless network, etc. These institutes are engaged in strengthening the communication between inner- and inter-industrial technology and building a technical and communicational platform for the industry. Via internal communication, Tencent will ensure smooth communication among basic technicians, applied technicians, and production technicians to make basic technology research become the terminal applications beneficial to the user. Via external communication, Tencent research institutes will build a regular communication mechanism for enterprise and national research institutes to improve the overall level of development and application of the Chinese Internet industry.

4. Summary

Facing the future, insisting on independent innovation, and establishing a national brand is Tencent's long-term development plan. It has a number of patent applications in instant messaging, e-commerce, online payment, search engine, information security, games, etc., and has invested hundreds of millions of yuan in Beijing, Shanghai, and Shenzhen to set up the first group of Chinese Internet research institutes—Tencent institute. The institute is engaged in independent R&D on basic Internet core technology. Tencent's independent innovation has entered the chain of enterprise development, operation, and sales. Tencent is gradually moving on along the road towards independent innovation with the development of the national industry.

Enterprise Case 5

Xunlei—A dominator and pioneer in the downloading engine industry

In 2002, Xunlei was founded in Silicon Valley in the US by Zou Shenglong and Cheng Hao. In 2003, they established Sandai Technologies Co., Ltd. in Shenzhen. Later, the company was renamed Xunlei Networking Technologies Co., Ltd. It focuses on providing the best multimedia resource data transmission services to the worldwide Internet. With core technology in the field of downloading, Xunlei has created a platform containing downloading, online video playing, games, and lots of other products. It has become one of the best network platforms with the most extensive coverage and the most users in China. As the biggest downloading service provider in China, Xunlei serves over 10 million downloadings from dozens of countries every day. Along with the popularization of the Internet broadband in China, it has become the byword for high-speed downloading because of the "simple and fast" downloading experience it provides.

1. The history of Xunlei's development

1.1 The initial stage from 2003 to 2004—Building a technology platform

Xunlei is one of the models that overseas returnees start up their own business in their motherland. The Internet is mainly applied in searching, portal, IM (instant messaging), and downloading software. Apart from the downloading software industry, there were already dominators in the other three fields. In the searching field, there were Google and Baidu; in the portal area, there were Sina and Sohu; in the IM field, there were QQ and MSN. After an analysis of the market, Zou Shenglong and Cheng Hao decided to develop downloading software. In 2003, Xunlei's first downloading software failed to get popularized in the market because it did not work efficiently. Later, Xunlei focused on the research on downloading technology and developed P2SP. The difference between P2SP and P2P is that the data of P2P comes solely from the users' PCs, while that of P2SP comes from the Internet servers. Comparatively speaking, P2SP could save more broadband resources and its downloading speed is 7–10 times faster than that of common software; in addition, it can reduce the percentage of dead links. Moreover, it supports multi-node HTTP, different downloading speeds, and can analyze intelligently which node uploads fastest. Compared with traditional Internet downloading, the speed of Xunlei saves human beings a lot of time—792 years every day. Its P2SP technology provides a great technical platform for its future development.

1.2 The business expansion stage from 2005 to 2006

Since 2005, after several upgrades and improvement of downloading engine software and service, Xunlei has launched new products like Thunder, Mini Thunder, Web Thunder, etc. and has thus entered a rapid growth stage. Xunlei's main business has expanded to downloading and searching portals of resources, blog services, Internet digital entertainment, etc.

1.2.1 Expansion to downloading and searching portals

Since 2005, Xunlei had launched a resource searching engine—Thunder Online. With Thunder Online, users could download videos, software, games, etc. It thus had become a backstage downloading tool of web portals. With the technological advantage of its downloading software, Xunlei launched a whole new Web Thunder. This product allowed users to search and download in the same interface. As a new resource searching portal, Xunlei took advantage of the fact that Google and Baidu had not done their job in the searching segment. It classifies different resources according to users' demand, and makes web searching and downloading more targeted. Unlike searching portals such as Google and Baidu, Xunlei considered more downloading factors in resource searching. Later, it launched Thunder Tingting, Mini Thunder, and InMedia Stream Media Advertising one after another. The web portal gave Xunlei a larger platform to integrate its existing resources. All of this was very useful to its growth.

1.2.2 The entrance into Internet digital entertainment

The multimedia digital entertainment publishing service had gradually become one of Xunlei's directions in order to make profit. On the one hand, it adopted Microsoft's latest DRM (digital rights management) system. This system provided a security strategy that prevents the possibility of exchanging licenses among different PCs. On the other hand, by strengthening its cooperation with film and records companies such as CETV, Hunan TV, Phoenix TV, etc., Xunlei got the exclusive internet distribution rights for some films and records. It only cost 4 yuan to watch a film with legal copyright on Xunlei's website. And it had over 100 million users. If only 10% of the users chose to watch a film on its website, the film producer will receive a considerable amount of profit. Xunlei's Internet digital entertainment publishing service provides a new channel for the publication of films with legal copyright.

1.3 Large-scale operation: Since 2007

Since 2007, Xunlei has gradually entered a period of large-scale operation. Its business involves Internet video, online games, value-added services for membership, digital publications, etc. It began to show up prominently in these areas. In 2007, it first released Thunder Kankan and innovatively realized cross protocol downloading. In 2008, it released Software Assistant, Security Assistant, and Thunder Games. In 2009, it executed its service order mode based on cloud computing and released Thunder 5.9, membership system, family digital entertainment products, web game, etc. In 2010, Thunder 7 was released with its new logo, and this opened a whole new era of Internet downloading. Xunlei also started strategic cooperation with Huayi Brothers to build an online video base for Chinese films. In 2011, Thunder Kankan became the first online video site that realized HD throughout the Internet and was also released for iPad, and Thunder for Mac was released in the same year. According to the data provided by iResearch, Xunlei served approximately 138 million downloadings every day in 2010 and the number of its paying memberships surpassed the 1 million mark. (In the first quarter of 2016, the paying membership reached 5.13 million.) In February of 2011, Thunder owned 78.7% of the downloading software market share in China. It has users in over 50 countries and regions around the world. Xunlei has gradually developed into a multinational corporation.

2. The enlightenment of Xunlei's innovation

2.1 Innovation-inserting strategy that gives wings to Xunlei

Facing the fierce competition in the Internet field, Xunlei came up with an unconventional approach: Applying an innovation-inserting strategy. Along with the innovation mode for R&D–manufacturing–marketing–users, Xunlei has found its own unique innovation method. The so-called innovation-inserting strategy means opening a new competitive area through subversive technical innovation and a unique road towards development and towards becoming the leader of this competitive field. Insertion means opportunity. On the basis of improving P2P, Xunlei developed its unique technology of P2SP and opened a new field in engine downloading. Consequently, it quickly became the dominator in the engine downloading industry. Furthermore, after analyzing the fields where there were already Internet dominators, Xunlei expanded its business into different fields according to its own characteristics. It is this innovation-inserting strategy that gives Xunlei a place in the fierce market competition. It is true that innovation inserting is the beginning of a successful enterprise, and continuous leading innovation is the key to maintaining the success.

2.2 Speed—the life of Xunlei, time—the wealth of Xunlei

With economic globalization nowadays, "time" is the first cost and "speed" is the key to success. Only with "leading speed" can an enterprise grasp opportunities. Searching needs speed, and downloading needs more. Speed depends on technology and resources. Xunlei has about 150 million users and meets their need for speedy downloading due to its leading technology. It could save the whole world 1,100 years every day, and it has become the byword for "top speed." As the most popular downloading software, Xunlei's top speed of downloading mainly depends on its own multimedia search engine downloading technology, based on binary, and on its intelligent transmission technology, which could efficiently integrate the downloading resources on the Internet and shape a unique Xunlei network. In this way, it is possible for users to download the resources they want as quickly as possible. It thus greatly meets the demands of the Internet users. The key to Xunlei's continuous development is to increase its download speed and save users as much time as possible.

2.3 Sustainable independent innovation—the unchangeable principle that keeps Xunlei alive forever

Once-only innovation cannot guarantee the sustainable growth of an enterprise. Continuous innovation can. As a company that focuses on engine downloading technology, Xunle's key to innovation is to make sure the downloading speed, security, and content. In terms of speed, Xunlei invests more than 80 million yuan in R&D and operation every year in order to adapt to the rapid growth of the market. Through continuous reform of its technology, Xunlei improves traditional downloading technology and downloading speed. In terms of security, it makes an effort to ensure the security for users while downloading. At present, 3% of the files downloaded from the Internet contain code, virus, and pop-up ads with different degrees of danger. Normally, a computer would suffer from virus twice to three times a month. The degree of security while downloading is very low. To solve the security problem, Xunlei participates in the Internet Downloading Security Alliance, and cooperates with domestic and international mainstream anti-virus and anti-hooligan firms. In terms of webpage downloading anti-virus, file downloading anti-virus, and the repair of the users' computers, Xunlei actively cooperates with Rising, Kaspersky, Kingsoft, Norton, etc., and has gradually discovered and established a sample database containing over 10 million harmful files. Xunlei has also cooperated with anti-virus firms in order to search for and kill viruses. In terms of content supervision and distribution, Xunlei does its best to satisfy its users through continuous improvement. To strengthen the company's vitality, Xunlei keeps on carrying out continuous innovation with sustainable development.

3. Summary

For a long time, "downloading" had been the ultimate application of Internet media development. When "downloading" is more than downloading and changes from client to "downloading engine," the downloading industry becomes the epitome of Internet applications. The rich content on the Internet made the "Information Era" replaced by the "Information Economy Era." The value of the latter is that information itself is not scarce and the promotion of value comes from information screening, processing, purifying, and obtaining. This process reminds us that when the content of the Internet becomes abundant, the downloading engine is more important than at any time in history. After the Chinese Internet industry has gone through the different stages of chill, divergence, boom, and diversity, the greatest demand of users will undoubtedly become the most direct navigation, which leads the new direction of Internet value. Only when the users' and the enterprise's values get unified, can an Internet enterprise achieve long-term development.

Enterprise Case 6

Mindray—The leader of the Chinese medical equipment industry

Mindray Medical International Limited (Mindray) is not only a leading Chinese manufacturer of high-tech medical devices, but also one of the global leaders in medical device innovation. Since its foundation in 1991, Mindray has always devoted its efforts to the development and manufacture of medical devices for clinical care. Its products cover four major fields: Life information and support, clinical examination and reagent, digital ultrasound, and radiography. Mindray takes electronic medical products with a perfect balance of performance and price to every corner of the world.

1. Mindray's growth

1.1 Major stages of Mindray's growth

Mindray's growth has gone through three major stages.

1.1.1 The stage of foundation and exploration: 1991–1997

At the beginning of its foundation, due to the lack of funds, Mindray was the agent of medical equipment for some well-known foreign brands like GE and Hitachi. The experience as an agent not only allowed Mindray to accumulate funds for business development, but also laid a solid foundation for the enterprise to understand market demands and to develop further in the future. With the development of the enterprise, Mindray's leadership team was facing problems of slow response of foreign companies and high prices of components. To promote sustainable development, Mindray began to concentrate on independent R&D and increased investment in R&D, production operation, and marketing promotion. Beginning from patient monitoring equipment which has a low threshold, less input, fast output, and a large market, Mindray chose an optimal cost-effective path and increased efforts on its marketing expansion. Consequently, Mindray was recognized by the market and gradually established a solid competitive advantage in the industry. In 1992, Mindray developed the first Chinese single parameter oxygen saturation monitor; due to its immature technology, quality, and marketing, the product did not get market acceptance at the beginning. In 1995, Mindray's product gradually gained recognition by the market, and Mindray got the ISO9001 certification of the German Tüv PS.

1.1.2 The stage of independent R&D and domestic leader: 1997–2002

Through the introduction of foreign venture capital, Mindray promoted its internationalization process in management, market, technology, and capital, and its financial indicators grew steadily. By the end of 2001, Mindray's life-care products ranked third in the Chinese market share after Agilent and GE and were sold to more than 40 countries and regions.

1.1.3 The diversified stage of internationalization and specialization: Since 2003

Mindray seized the opportunity to develop the path towards internationalization and specialization, and it gradually expanded its business to more than 190 countries and regions. Mindray was listed on the New York Stock Exchange (NYSE) successfully in September 2006. Mindray entered four product fields: Life information and support, clinical examination and pilot production, digital ultrasound, and radiography, and its sales maintained an annual increase of 50%, and the company launched new products, leading a trend in the industry. Mindray gradually developed into a leader in the domestic medical device industry and was extremely competitive in the international market. In 2010, it was recognized with the "2010 European Patient Monitoring Excellence in Globalization Award" and the "2010 Innovation Excellence Award" by the worldwide authoritative market research firm Frost & Sullivan. Mindray's success lies in the company's behavior in insisting on independent innovation and adopting an appropriate strategy in each of the different stages of its growth.

1.2 Learning from Mindray's growth

1.2.1 Independent R&D, the vitality of Mindray

Since 1992, when it developed the first oxygen saturation monitor with independent IPR, Mindray has introduced nearly 10 kinds of monitors, including the first set of "anesthetic gas monitors" in China. Through independent R&D and continuous innovation, Mindray has firmly occupied the leading position in three series of products: High-end patient monitors, digital ultrasound (B-ultrasound), and clinical examinations and reagents. Its cost savings and efficiencies brought about by complete independent IPR guarantees its products sell well and its market expand to overseas countries. According to customs statistics, Mindray's products account for more than 60% of the high-end medical devices exported overseas from China.

Currently, Mindray is setting up research centers in Shenzhen and Beijing, and has established specialized laboratories and EMC testing laboratories at an advanced international level; this gives shape to its strong technological R&D and innovation system. In 2002, the "National Medical Diagnostic Instrument Engineering Technology Research Center" approved by the Ministry of Science and Technology, became the only national engineering technology research center in Shenzhen, which guided and enhanced the R&D process of Chinese medical devices by new R&D mechanisms. Successful experiences in R&D project management and a rigorous scientific R&D management system improve the probability of Mindray's successful innovation. As a result, Mindray has gradually established itself in the leading position in China.

1.2.2 Capital that makes Mindray realize its value

From Walden's venture capital to its successful IPO on the NYSE, successful capital operations have contributed to Mindray's overall success. In 1997, Mindray got a $2 million investment from Walden, which sped up the R&D project and helped to establish a market network. Walden not only provided Mindray with funds, but also brought advanced management experience to perfect the corporate governance structure. In 1999, Walden invested $6 million in three venture capital firms: Softbank, JAIC, and Asia Shengchang. In 2000, although Mindray passed the relevant procedures of the China Securities Regulatory Commission (CSRC) and the Hong Kong Stock Exchange (HKEX), the IPO was unsuccessful on the Hong Kong Growth Enterprises Market (GEM). In 2001, Mindray restructured itself to a foreign investment company limited. In the same year, it tried to go public, but met the Wall Street stock exchange crash. The terrible capital market finally forced Mindray to give up going public. In 2006, Mindray was successfully listed on the NYSE as the first Chinese medical equipment enterprise. Mindray's IPO improved its brand awareness and reliability on overseas markets, and its competitiveness as well, which laid a solid foundation for further expansion on the overseas market.

1.2.3 Entering overseas markets to give Mindray a chance to show its ability

Entering overseas markets in 2000 was a milestone in Mindray's history. Currently, the capacity of the foreign market is much larger than that of the domestic market. American and European markets account for 70% of the amount of global medical device usage, while the Chinese market is only about 7%. Taking the patient monitor as an example, every 100 beds are generally equipped with 40 monitors in medical institutions of developed countries; in contrast, the number in China is only 20 in the best medical institutions. The number in the majority of medical institutions is only one or two. Entering overseas markets and improving its market share of those markets is the key to Mindray's sustainable development.

Concerning the foreign market, Mindray first got the European system certification and went onto the European market, especially in developed countries, such as Britain and Germany. Depending on its European agents and the cost-effective advantage of its products, Mindray gradually occupied some of the foreign market share. In 2006, Mindray's sales revenue reached about 1.5 billion yuan, of which the domestic market accounted for 51% and the international market accounted for 49%. Compared with 2005, the sales on the domestic market grew by 24.3% and those on the international market grew by 62.9%. Regarding the patient monitoring products, Mindray has accounted for 63% of the total domestic export volume. With Mindray's successful listing in New York, discrimination from the international market towards Mindray decreased and the number of agents gradually increased, which promoted the development of Mindray to some extent. In 2010, Mindray's revenue was $700 million, 58% of which came from the foreign market and 42% from the domestic market. According to a long-term analysis, the foreign market is still Mindray's major market.

2. Mindray's all-around independent innovation

2.1 Technological innovation that starts from high-end

Being guided by technological breakthroughs and starting from high-end technology are keys to Mindray's success. Because of the high technological barrier in the medical device industry, it is difficult to get a foothold on the market without a technological advantage. In 1992, when lots of medical device companies were still satisfied with the development process of trade, industry, and technology, Mindray invested all of its profits in the development of new products. Up to now, Mindray has always followed the rule that 10% of the sales revenue should be retained for R&D spending. Moreover, Mindray has established its R&D management platform matched with the first-class international level and launched more than 80 kinds of products with independent IPR. By the end of 2010, it had submitted about 1,100 domestic and international patent applications, more than 550 of which were authorized.

The products developed by Mindray's independent R&D with independent IPR, such as the portable digital ultrasound diagnostic system, the BS-300 automatic biochemical analyzer, and the transcranial Doppler CBF diagnostic apparatus, not only won nearly 20 "Chinese firsts," but also obtained a good market. It is starting from the high-end that makes Mindray grow rapidly in a short time. In addition, Mindray attaches great importance to the study of key technology in advance, the study of technical regulations, and the investment in R&D, and it has established laboratories at an advanced international level. In terms of internationalization, Mindray has begun to carry out international product R&D. It has also established a research institute in North America. By cooperating with multinational corporations in carrying out R&D, Mindray has mastered the product requirement of GE and FD. At present, Mindray has shaped the industry–university–medicine cooperation mechanism, where the enterprise is the subject and the universities, the research institutes, and multinational companies are important participants. Through its cooperation with hospitals, Mindray has built a clinical base to provide a good platform for independent innovation.

2.2 Management innovation that wins out over multinational rivals

Management innovation is the guarantee for enterprise technological innovation. The focuses of Mindray's management innovation are on the innovations in its talent training system, patent protection, and customer service system. The source of Mindray's technological innovation is talented people, and an excellent technology development team is the foundation of Mindrays' technical ability. As to its talent training system, taking "employee's lifelong learning" as an eternal principle of talent cultivation, Mindray provides training opportunities and internships for new staff, and establishes a sound system of staff training and career development. Its innovative talent training system increases the loyalty of technical talents, avoids the loss of high-end talents, and stimulates the enthusiasm and innovativeness of researchers to some extent, all of which lays a solid foundation for independent innovation.

While Mindray encourages talent management innovation, it also pays attention to IPR in the company's development strategy. Competition is fierce in the medical device industry, so intellectual property means respect for the developers' achievements and a protection for the interests of the enterprise. Therefore, a normative patent management system has been set up by Mindray. From the search for a patent novelty before the project to its later maintenance, Mindray has a strict management process. On the premise of respecting the IPR of others and using patent technology and relevant literature, Mindray uses the system to protect its IPR and then form its own unique market competitive advantages. Patent protection is the umbrella for enterprise technology and innovation. An effective patent protection system is an important support to the independent innovation system. With regards to the customers' service system, Mindray follows a 24-hour customer response system, which shortens the response time from the recognized two weeks to 24 hours, and overturns the original multi-level distribution mode established by multinational enterprises in China. The primary distribution system established by Mindray greatly shortens the distance between the company and its customers, reduces costs, and leaves some benefit space for customers. Mindray's management innovation provides a strong guarantee for its development in the areas of technology and technology commercialization.

2.3 Innovation being a continuous feature

Mindray would dedicate 10% of its sales revenue to R&D investment every year, and continuous R&D investment is the basis of Mindray's continuous innovation. In addition to a constant increase in R&D investment, Mindray has strict rules about the number of new products put onto the market every year and about the contribution rate of each new product with regards to corporate profits. There is an assessment quarterly, and the salary of R&D personnel is tied to their sales. 40% of the salary of R&D personnel is decided by the market and the selling conditions. Mindray has gradually formed a set of complete systems, which starts from technology and goes on to the market, and finally comes back to technology. Innovation has become an obvious culture of Mindray's.

3. The enlightenment from Mindray's innovation

3.1 Insisting on "independent innovation"

While adhering to independent innovation, Mindray gradually makes it become the enterprise's tradition and concept, and the driving force for further independent innovation. Considering investment in innovation, Mindray dedicates 10% of its sales revenue to R&D spending every year and to its more than 800 R&D personnel; as for innovation output, Mindray launches three kinds of new products per year. Insisting on independent innovation and taking innovation as the enterprise's concept and culture maintain the company's long-term development.

3.2 Inputting in different "baskets"

The medical device industry involves many branches, including software, automation, mechanical, biochemistry, medicine, and polymer materials, so it is a typical interdisciplinary industry. Once developed successfully, the production line can accumulate capability for expansion. Starting from patient monitoring products, Mindray has developed a platform of software and hardware with complete independent IPR and basic parameter modules, which become an important foundation for further technological innovation. On the basis of this technology platform, Mindray speeds up the development of its automatic biochemical analyzer, hematology analyzer, and other products in a shorter time, achieving a huge success in many different fields.

3.3 Relying on core technology to enter the international market

Laws and regulations constitute the hardest problem to solve for enterprises that want to enter the international market. The key to stepping into the international market is to get the certifications of CE and FDA; and having products with core technology is necessary in order to obtain these certifications. At present, Mindray enters the foreign market through its core technology products; and its revenue from the foreign market has exceeded 50%. Consequently, Mindray has gradually become an export-driven and technology-oriented enterprise.

3.4 The effective capital operation fundamental for its success

During its difficult period, Mindray got venture capital in time. The American Walden provided capital for Mindray twice. With the development of Mindray, shareholding reform and IPO provide sufficient funds for its subsequent development.

4. Summary

Digital medical device is the short board of the industrial development in China, yet the market gap is very big. China needs enterprises like Mindray to be the core force of the medical instrument industry. After going public successfully, Mindray has been developing constantly, driven by the strong innovation engine, thus creating a brighter prospect for the Chinese medical device industry.

Enterprise Case 7

CHINT—From a Wenzhou manufacturer to a world-class brand

1. The history of CHINT's development

CHINT was founded in July 1984 and consists of 8 specialized companies, over 2,000 domestic sales branches and distributors, and more than 50 overseas distributors. Its products cover high and low-voltage electrical apparatus, power transmission and distribution, meters and instruments, building and construction electrical appliances, auto parts, automation, and solar products. Its products are sold well in more than 90 countries and regions. The history of CHINT's development is not only the entrepreneurship history of a private enterprise, but also the history of the growth of a Chinese independent brand. After more than thirty years of hard work, CHINT stands out from among thousands of low-voltage electrical appliance enterprises in Wenzhou of Zhejiang. Now, its size already ranks first in Asia, and it is gradually moving towards internationalization.

1.1 The stage of start-up and accumulation (1984–1990)

Yueqing County Qiujing Switch Factory (the predecessor of CHINT) was founded in July 1984. At the beginning of its foundation, CHINT had a workshop with a surface area of only 50 square meters, eight employees, and capital of 50,000 yuan. Its annual output value was less than 10,000 yuan. After seven years of hard work, its annual output value reached 10.21 million yuan, and its total assets reached 2 million yuan. In line with the spirit of "Always striving for excellence," CHINT relied on quality and credit to survive. It completed its primitive accumulation of capital, laying the foundation for its following development.

1.2 The stage of targeted development (1991–1993)

In 1991, Nan Cunhui borrowed $150,000 and established Wenzhou CHINT Electrical Appliances Co., Ltd. After a period of development, Nan Cunhui persuaded his brother Nan Cunfei, his nephew Zhu Xinmin, his brother-in-law Wu Bingchi, and another relative Lin Liming to join CHINT Company and become shareholders, which completed the basic construction of CHINT. At that time, Nan Cunhui's personal share was diluted from 100% to 60%, and the remaining 40% was held by the other family members. This was the first stock dilution. After that, Nan Cunhui cooperated with an American businessman Huang Liyi in establishing a joint venture, which was called Sino-US Wenzhou CHINT Electrical Appliances Co., Ltd. They imported advanced technology and equipment from abroad and made full use of the preferential policies of the joint venture to improve its strength; moreover, they decided on the direction of electric professionalization. The landmark event was the completion of CHINT's head office building in September 1993. CHINT put forward the slogan of "Creating a new image of Wenzhou electrical appliances."

1.3 The stage of expansion and group operation (1994–1996)

In 1994, Nan Cunhui started the second revolution of property rights. Through selling shares of CHINT and holding shares, participating, or investing in other businesses, he completed the M&A of 48 enterprises. In February of that year, Wenzhou CHINT Group was established, the number of whose shareholders had sharply increased to 40, and the personal share of Nan Cunhui decreased to around 40%. CHINT integrated a large number of social resources, and its net assets increased from about 4 million to 50 million yuan. CHINT had realized the goal of scale economy, and Nan Cunhui's personal wealth had also increased by nearly 20 times within 3 years. During this period, taking capital as the link, market as the guidance, product as the lead, and brand as the center, CHINT moved on along the road of scale economy, leading to the rapid development and expansion of the enterprise through horizontal integration.

1.4 The stage of the transition from the Group to a holding (group) company (1997–2002)

In July 1997, the group's first standard limited liability company, Zhejiang CHINT Electrical Appliances Co., Ltd, was approved by the people's government of Zhejiang Province. In 1998, Nan Cunhui gradually carried out an equity distribution system in CHINT Electrical Appliances Co., Ltd., which was the core enterprise of CHINT Group. The system meant some elements can become shares, including management, technology, and operation, which attracted dozens of knowledge-type shareholders. The number of CHINT's core shareholders was thus expanded to 118, including the original investors, the shareholders converting from subsidiaries' owners, and the second-level shareholders, who were part of CHINT's technical, management, and marketing personnel. After that, CHINT conducted a demutualization reform, cut down overlapping institutions, and perfected "Three Boards." At that time, Nan Cunhui's personal share dropped to 20% and several other senior entrepreneurs, including

Nan Cunfei and Zhu Xinmin who held 7%–10% of the total shares. At that highest decision-making level, the proportion of Nan's family members was less than 1/3 and, among 118 core shareholders, they only accounted for about 20%.

In over ten years, CHINT had achieved great success in optimizing the structure of property rights by stock dilution for three times, so that the total assets of CHINT had increased to 3.1 billion yuan from the initial 2 million yuan. And CHINT owned six core subsidiaries and more than 50 branches. During this period, CHINT carried out a shareholding reform in its affiliated companies. By building a limited liability company and a company limited by shares as well as developing the enterprise in accordance with the requirements of a modern enterprise system, CHINT had moved forward towards a global enterprise.

1.5 Enhancing and spanning stage (2003–now)

CHINT set the goal of "creating a world-class electrics manufacturing enterprise," and has gradually shaped the "Yangtze River Delta Layout." In this layout, Wenzhou is the manufacturing base of low-voltage appliances, meters and instruments, and building and construction electrical appliances; Shanghai, the base for high-voltage power transmission and distribution equipment; Jiaxing, the base for supporting equipment of power transmission and distribution; and Hangzhou, the base for industrial automation and PV power production. Independent innovation has become the theme of the enterprise's development. The Group boasts more than 1,000 Chinese and international patents, and has led and participated in the formulation of more than 120 industrial standards. In January 2010, Zhejiang CHINT Electrical Appliances Co., Ltd. successfully launched its IPO at the Shanghai Stock Exchange (SSE) of China.

2. The developmental trajectory from independent innovation to independent brand

2.1 Shedding the label of imitative "small workshop"

At the end of the 1980s and the beginning of the 1990s, many areas boycotted Wenzhou low-voltage electrical products because of the fake commodities and the poor quality. The signs saying "Wenzhou electric salesmen are not allowed to come in" were hung everywhere. Six national ministries directly appointed a working group, stationed in Yueqing of Wenzhou, with the purpose of focusing on removing the fake and shoddy products. At that time, thousands of low-voltage electrical factories almost disappeared overnight, but CHINT survived because of its excellent product quality, while its competitors were under severe attack. From then on, CHINT has shed the label of imitative "small workshop."

2.2 Establishing an independent brand through product quality

As the predecessor of CHINT, Qiujing Switch Factory attached great importance to product quality at the very beginning of its establishment and insisted on building its own brand. It was the first factory to obtain a production license in Yueqing County and became one of the survivors during the counterfeit boycott, and was chosen by the governments of Yueqing City and Liushi Town as the key supporting company. Later, CHINT achieved its first expansion by using its own brand. In 1990, the enterprises facing bankruptcy in Yueqing County were merged into CHINT, and there were more than 40 companies joining CHINT Group within 4 years. CHINT's net assets rose from less than 4 million yuan to 50 million yuan. Facing the fierce competition on the market, CHINT gradually realized the importance of brand and product quality. It would choose to sacrifice 1 billion yuan of its output value, rather than allow a poor quality product to leave the factory. Due to its quality consciousness, CHINT stood out among many similar enterprises in Liushi Town, which is called the land of Chinese low-voltage electric appliances. Since 1995, every May has been CHINT's "quality month." Since 2002, CHINT has implemented the system of "one-vote veto" and gradually established a set of effective quality control systems. Besides, there is a quality-inspection team with hundreds of people, a "president-inspection" system, and "the chairman line." CHINT has received a lot of certifications, including the ISO9001 quality management systems, the ISO14001 environmental management systems, the OHSAS18001 occupational health and safety management systems, the international CB safety certification, the American UL certification, and the German VFE certification; and it also received the first China Compulsory Certification (CCC) of the low-voltage electrical industry. In 2001, CHINT became the first private enterprise of the low-voltage electrical apparatus industry to win the National Quality Management Award, which is the highest national award in the quality field. This prize is the most authoritative approval of its quality management.

2.3 Consolidating an independent brand by science and technological innovation

Quality alone provides little potential for the development of a brand; while the core technology can produce enduring force. Only by insisting on innovation and mastering core technology, is it possible to create a world brand. 3% to 5% of CHINT's sales revenue is invested in the S&T development each year. CHINT has invested heavily in order to import world-class R&D equipment, set up first-class international physical and chemical testing centers, metering centers, low-voltage electric testing centers, and a high and low-voltage electric R&D base. CHINT elaborately constructed three-level R&D institutions—a national level technology development center in Shanghai, R&D companies in specialized companies, and R&D departments in all enterprises held by CHINT. Besides, global R&D centers were set up in Shanghai, Prague, and Los Angeles. Cooperating with some well-known domestic universities and research institutes, CHINT has formed an open information network and technology development system.

In addition, CHINT has recruited more than 1,000 intermediate and senior personnel across the country, has held S&T conferences every two years, and has rewarded departments and individuals who have made contributions. Some outstanding people will be hired to become shareholders in order to inspire their enthusiasm and creativity. Through continuous independent innovation, up to 2015, CHINT had successively developed over 100 series and 200 basic models of new products with independent IPR, including high and low-voltage electrical appliances, complete sets of equipment, optical fiber distribution equipment, and intelligent instruments. It is through continuous innovation and new products development that CHINT has consolidated its position as an independent brand.

2.4 Taking IPR as the amulet of the brand

Over the past 30 years, CHINT has encountered several lawsuits related to IPR; as a consequence, it has recognized the importance of competitive advantage and advanced technologies. Therefore, CHINT carries out a technology blockade on the one hand, and a "patent thicket" strategy on the other hand. The so-called "patent thicket" means taking advantage of the frontier technology to create a huge patent database and taking over patent rights, which will bring difficulties for the later R&D companies and force them either to bypass the existing technologies, or to pay huge fees to the forerunners in order to acquire the right to use their technologies. In order to safeguard its independent IPR, CHINT has established a special primary department called "intellectual property department," which is independent from its R&D institution and controlled directly by the Group. Its main function is to search for patents in related global fields. According to the database of the intellectual property department, the R&D department works on filling the innovation gaps in these fields on the basis of understanding the existing patents, so as to extend the patent reserve.

3. Selecting the trajectory for CHINT's independent innovation

From its technological innovation strategy, it can be seen that CHINT takes the trajectory of technological innovation considering both internal and external factors, and focuses on both basic and applied research.

3.1 A clear hierarchical structure of technological innovation

CHINT's technological innovation has a strategic significance, including a strategic plan and tactical measures, and it contains three aspects: The first is the "catching up and surpassing level." A research institute was established in Silicon Valley in the US in 1999. The institute was mainly engaged in the research of basic and frontier global S&T as well as the latest industrial technology. The second is the "updating level." CHINT considers the scientific research institute in Shanghai as the developing center, aiming at serving the industrial group directly and providing a full range of technical support to related industries. The third is the "improving level." The technical departments and technology sectors of each branch take responsibility for the designing and improving of the process for various existing products as well as improving their quality.

3.2 Mastering core technology in all aspects

In order to master core technology and to avoid deviating from its strategic direction, CHINT has done the following: Regarding technology, CHINT has established the principle of "developing the market according to our own characteristics" and insisted on the "three predictions" strategy—the product's technical life, its market capacity, and its economic benefits concerning the position of technology in product development. With regards to the development of new products, the company has established the principle of "total IPR" and insisted on "thinking twice before making a decision." Regarding talents, it has recruited domestic and foreign technicians in every field so as to maintain the sustainability of innovation. CHINT constantly cultivates enterprise technicians and makes extensive use of social scientific research.

3.3 Realizing "two transformations" and insisting on "two combinations"

CHINT has realized two transformations in product and production operation. The first one is the transformation from mechanization to electronization, intelligence, modularization, and miniaturization. The second, from labor-intensive to tech-intensive. Over the past 30 years, CHINT has always insisted on the trajectory of combining independence and openness as well as introduction and innovation. On the one hand, while continuously expanding its market and opening up to the outside, CHINT has introduced advanced overseas technology; on the other hand, it keeps cultivating its innovation ability on the basis of adopting and implementing the advanced technology from abroad. The stronger the independent innovation ability is, the more solid the basis of the independent brand is. And it will be much stronger through negotiation and cooperation with multinational enterprises. Openness and independence as well as introducing technology and improving independent innovation ability are complementary. At present, CHINT has adopted the technology of CAD and CAM comprehensively, it has implemented the project of CIMS, computer integrated manufacturing system, and equipped itself with an advanced CNC tooling center as well as flexible manufacturing equipment. Consequently, CHINT has shaped a multi-level and open-tech development network, which is centered on group technological development with professional technology as the subject. CHINT has also established an "innovation chain" which integrates scientific research, teaching, training, and development. CHINT has successively developed more than 200 new products with independent IPR, which have the features of intelligence, miniaturization, modularization, and communicable fieldbus. Among these new products, up to 2007, 62 had passed provincial certification, and 18 had reached the advanced national and international level. Also, three had become the National Torch Plan projects, and obtained more than 100 national patents. Moreover, four of CHINT's products received awards of China Top Brand in 2006, including low-voltage switchgear, watt-hour meter, air-operated circuit breaker (ACB), and molded case circuit breaker (MCCB). CHINT brand covers three leading industries: Complete sets of equipment, low-voltage electrical appliances, and instruments and meters.

3.4 Boosting "three innovations"

System innovation and management innovation guarantee technological innovation. While implementing technological innovation, CHINT conducts system and management innovation to promote its sustainable development.

3.4.1 System innovation

CHINT has gone through four reforms and optimizations of its property rights system in its entrepreneurship. The first time was in 1991, when CHINT changed from partnership to incorporation; the second was the attempt to change from incorporation to group in 1994; the third was the transformation from group to joint-stock in 1996; and the fourth was the further standardization of the joint-stock company system at the end of 2000. These four reforms of the property rights system were carried out on the premise of different external environmental and internal managerial needs. The first and the second belonged to CHINT's first period of entrepreneurship; the third, the second period of entrepreneurship; and the fourth, the prelude of its third entrepreneurship.

In terms of system innovation, CHINT took the lead in taking an important step that shocked many private enterprises, namely breaking the single ownership structure fundamentally and improving the foundation of its management system, including the introduction of family capital to dilute personal capital and the introduction of social capital to dilute family capital. Through this kind of equity reform, CHINT has realized the proper separation of enterprise ownership and management rights, which leads to the diversification and socialization of property rights. In its daily operations, CHINT has established the check and balance system of three boards (board of shareholders, board of directors, and board of supervisors) and the separation of four rights (ownership rights, decision rights, management rights, and supervision rights). This system provides independent innovation with a sound incentive mechanism.

3.4.2 Management innovation

CHINT began to implement grouping operation from 1994 and then a corporate system of a holding group, where the group was divided into three levels: The sector of investment decision-making (which is controlled by the headquarter of the group and responsible for capital operation, human resources management, and other strategic issues), the sector of profit growth (including production companies and sales companies), and the sector of cost control (made up of production companies, factories, and workshops). It forms the enterprise's management system, which adopted the parent company as its basic framework, and also the decision-making mode with the combination of expert consulting, and democratic decision-making.

3.4.3 Cultural innovation

Under the conditions of an open market economy, enterprise culture has increasingly become the inner demand to display enterprise competitiveness and enhance the cohesive force of an enterprise; it is also the source of vitality for the development and progress of an enterprise. The cultural factor exists in an enterprise from the very first day of its foundation, and also influences and constructs different enterprises. Cultural competition is the essence of market competition that is the carrier of cultural competition. In the unique geographical and cultural background, and after over 30 years of experiment and development, CHINT was founded out of nothing and has formed a unique enterprise culture with its own characteristics. CHINT advocates the group values of "people-orientation" and "serving the society," and takes these values as the focal point of its enterprise culture; besides, it promotes the business mission of "harmony, modesty, practicability, and innovation." What is more, its enterprise concept is "creating a world-famous brand and serving the country through industrial development" and it integrates the staff's value with the goal of enterprise development. In addition, CHINT has also established academic institutes and education training centers in order to become a "learning organization" from the top to the bottom; in this way, CHINT can promote its enterprise culture and cohesive force while heightening the vitality of its personnel. In the competitive global environment, the Group has put forward the principle of "people-orientation, civilization, quality, and image." CHINT is striving to develop from a well-known Chinese trademark to a global brand.

4. Summary

The key point in the development of CHINT is sticking to independent innovation and independent brand. In order to improve its international competitiveness, an enterprise must cultivate its own core technology on the basis of the introduction, reference, and adoption of advanced international technologies. Before CHINT can complete its transformation from a "grassroot enterprise" to a world-class one, it still needs to make efforts in four aspects: The first is to adhere to independent innovation and change from extensive to intensive operating; the second is to continuously optimize its industrial structure and transform from a traditional enterprise to an energy-saving and environment-friendly enterprise; the third is to actively change its mode of operation and switch from product management to capital management; the fourth is to actively implement the "going out" strategy and transfer from a Wenzhou manufacturer to a world-class brand.

Enterprise Case 8

Nuctech—The clairvoyance of China

1. Company profile

Nuctech Company Limited (Nuctech) was founded in April of 1997, originating out of Tsinghua University. It specializes in the development, production, sales, and service of radiologic technology products. Its products are widely applied in customs, industrial manufacturing, national security, health care, environmental protection, food processing, etc.

Currently, its main products are the "Nuctech" series of cargo and vehicle inspection system. Cooperating with Tsinghua University, Nuctech focuses on high-tech industrialization and has developed the high-performance series of cargo and vehicle inspection system, air cargo and vehicle inspection system, electronic beam mail sterilization safety system, industrial NDT system, etc., which employ linear accelerators and represent the most advanced level in the world and a trend in the development of technology. Nuctech owns the professional devising and producing ability of the entire core technology. It can manufacture specialized inspection systems according to the demand of its clients, provide integrated solutions, and undertake auxiliary projects.

Its cargo and vehicle safety inspection systems can be divided into three types: The relocatable, the mobile, and the fixed inspection systems. These products are widely used in sea ports and border crossings. Nuctech owns independent IPR together with the whole core technology in the radiation application field. It can provide tailored and integrated solutions to the customers. Besides, it can also provide them with systematical training, project implementation, and after-sale maintenance services.

Based on advanced technology and excellent service, Nuctech has got a market share with absolute advantages in the international competition. Currently, its products have already been exported to over 140 countries and areas in Europe, America, Oceania, Asia, and Africa. The systems put into operation abroad work well with notable inspection results and win popularity and high evaluation from the customers. Besides, Nuctech has also built a good reputation and brand image both at home and abroad. In accordance with Tsinghua University's motto of "Self-discipline and Social Commitment," Nuctech makes developing S&T and serving the society its task, and becoming "No.1 at home and in the world" its goal. Nuctech strives to join forces with all sectors of society to make improvements together.

2. The analysis of Nuctech's trajectory towards independent innovation

Nuctech aims at independent innovation, integration innovation, and introduction–adoption and implementation–re-innovation. It owns independent IPR of the entire core technology. Setting its target on a healthy, stable, coordinative, and sustainable development, Nuctech insists on the "going out strategy" and focuses on coordinative development between the domestic and international markets. For many years, in accordance with Tsinghua University's motto of "Self-discipline and Social Commitment," Tongfang's spirit of "Responsibility, Exploration, and Surpassing" and the concept of "Let us make progress together," Nuctech follows the demand of the country and society to provide high-tech products.

2.1 The collaborative innovation of the university and the enterprise: Technology plus capital makes an incubator

The industrialization of technology is a common problem for universities, research institutes, and enterprises. Some of them may have good technology, but separated from the market; some may have good technology and there is a demand on the market, but they lack capital, so it is difficult to carry out industrialization. In addition, the lack of capital also discourages scientific researchers. However, Nuctech has successfully established the precedent that "technology plus capital makes an incubator." Tsinghua offers its scientific achievements to its enterprises including Tongfang and Ziguang, so that technologies can be transformed into products in order to step into the market. Nuctech was established based only on one of the technologies. In order to speed up the industrialization, Tsinghua transformed its scientific research system into an enterprise operation system and put the original task groups and new experts into Tongfang in order to incubate it to become an enterprise. This enterprise is responsible for industrializing the scientific achievements. Normative enterprise operation enables the most effective allocation of capital, technology, and talents in the shortest time span. Besides, it even makes the scientific research mode dominated by major transforms into the innovation mode dominated by products. Currently, Chinese venture capital (VC) system is not sound, and therefore such a mode is more suitable to the national conditions. Unlike the successful mode of Silicon Valley where most enterprises grow based on private scientific and technological achievements with the help of VC, the success of Nuctech is attributed to the great power behind it.

The success of Nuctech lies in finding a sustainable independent innovation trajectory—the collaborative innovation of the university and the enterprise which takes "technology plus capital" as the core. The collaborative innovation of the university and the enterprise not merely means "holding hands together," but also means finding a stable technological patron for the enterprise. On the one hand, the transformation of Tsinghua University's technological achievements into products through industrialization achieves an economic benefit and makes a contribution to the country. On the other hand, with the help of the technological research at Tsinghua University, the research pressure on Nuctech can be relieved so that it can focus on the industrialization of its products. Zheng Yongping, former director of the Office of Intellectual Property Rights Management at Tsinghua University, and his partners (2008) summarize the collaborative innovation of the university and the enterprise in four modes: The government-driven mode, the technology transfer mode, the cooperative development mode, and a joint R&D entity. The government-driven mode refers to establishing a company based on the patent technology of the university with governmental support, and Nuctech is a classic case.

2.2 Industrialization, the eternal motivation of independent innovation

Industrialization is the power source of enterprise independent innovation. Only with the economic benefit coming from industrialization and the transformation of technology into products, can the sustainable independent innovation of enterprise be guaranteed. Nuctech owns a complete industrialization system and a professional production base of 100,000 square meters for its container inspection system, the biggest in the world, as well as a modern production and testing devices. The production base is equipped with advanced production, installation, testing, and debugging systems. Coupled with scientific management, in accordance with the standard demand of the ISO9001 international quality assurance system, Nuctech conducts strict control to all links, including design, procurement, production, testing, installation, commissioning, debugging, and after-sale service, so as to guarantee a high quality of production.

The industrialization of Nuctech is special; it can customize and devise systems according to different areas and special demands of clients in order to provide a perfect customized solution for each client. For example, the products exported to Norway can withstand temperatures below −30℃ , while those to the United Arab Emirates can operate normally even above 50℃ . During its development, Nuctech has gradually developed the R&D trajectory of market demand–technological research–industrialization.

2.3 Independent IPR, the most important achievement of independent innovation

Intellectual property strategy is an important part of Nuctech's entire management strategy. The effective implementation of the strategy plays an important role of protection in the development of Nuctech. It promotes and guides the company to constantly maintain its technological innovative capabilities and core competitive advantages. Nuctech has built a well-developed and scientific framework for the IPR management system: A three-level framework for the patent management and operation system; a three-in-one talent framework, including technology, law, and management; a network for the layout of the framework for the IPR protection strategy, which is prompted by business secret, patent, trademark, and software copyright; a framework for the strategic decision-making mechanism that includes the market, the technologists, and the operating management. The effective operation of all of the above-mentioned systems and frameworks guarantees the healthy operation and development and constant enhancement of the competitiveness of Nuctech.

Nuctech owns complete independent IPR together with the core technology. In the radiation application sector, Nuctech has independently developed a large cargo and vehicle inspection system, an air cargo and vehicle inspection system, a railway cargo/vehicle inspection system, a mail sterilization system, a radioactive substances monitor system, a small items inspection system, a liquid inspection system, an industrial nondestructive inspection system, etc. Up to the end of 2013, its inspection products took up 70% of the global market share and its clients include over 140 countries and areas from Asia, Africa, Europe, Australia, and South America. The number of patent applications increases at a speed of 100% annually. By the end of 2015, Nuctech had already applied for over 2,800 patents both at home and abroad. The core technology of its products has already reached an advanced international level, and the "R&D and industrialization of mobile cargo inspection system series using the accelerator as the radiation source" won the first prize of the National Science and Technology Progress Award. "Relocatable cargo inspection system" won the Chinese Patent Golden Prize awarded by the State Intellectual Property Office and the World Intellectual Property Organization. Nuctech's brand influence is constantly improving and highly evaluated by clients both at home and abroad.

2.4 The ultimate goal of independent innovation: An independent brand

Forming an independent brand is one of the goals of enterprise innovation. The innovation of Nuctech forges its special brand image. Currently, its main product, cargo and vehicle inspection system, ranks among the top three in the world. The reason why Nuctech can stand out on the international market is its powerful strength. For a long time, Chinese products had a negative evaluation for its low-price and low-quality on the international market, and thus there had been a hard time for Nuctech initially. In order to make the world have a better understanding of Nuctech, it intensified its advertisement and invited customs officials from various countries to visit the production base for them to know the products and carry out field tests. Finally, Nuctech opened the market in Australia and obtained its first success in 2001. And at the Dubai port, the United Arab Emirates, Nuctech competed with American and German products and finally won due to its quality and service. Now, Nuctech has already signed contracts with dozens of countries, and has become a well-known international brand.

Considering Nuctech's multiple successes in competition, we can see that the key point of sustainable competitiveness of Nuctech relies much on its ability to customize its products. Since a cargo inspection system needs to be customized according to the demands of the customers, the degree of customization means the enterprises' design and production ability, which decides the core-competitiveness of the product. Nuctech has the core technology with independent IPR, a technical team with an excellent capability for customized design, and a manufacturing mechanism which is suitable for the customization of numerous varieties and small batches, and is able to meet the demands of the clients instantly. Through customization, Nuctech meets the new and rising demands of clients, and extends and deepens the influence of its brand. Through the various services of cargo inspection, Nuctech consolidates the influence of its brand.

If a good product or brand wants to stand out on the changing international market, it should not only have advanced technology, but also combine innovation technology with customized service. In the international competition, Nuctech not only carries out technological innovation, but also shapes an industrialization platform based on its core technology, from which Nuctech makes an effort to create unique "customization." As a black horse on the world cargo inspection equipment market, with its first-class technology, quality, after-sale service, and competitive prices, Nuctech will certainly become a renowned international brand.

2.5 Government procurement and support

The cargo inspection system as Nuctech's main product, relates to the national economy and the people's livelihood and involves large amount of capital, so government procurement becomes the main marketing channel. The Chinese government is used to supporting enterprise independent innovation by government procurement products, which gives great support to independent innovation technology, products, and industry. In January 1998, when the large cargo inspection system passed appraisal, the State Science and Technology Commission and the General Administration of Customs signed the Framework Concerning the H986 Project. In May and June 1999, Nutech's MT1213 mobile cargo/vehicle inspection system and its MB1215 relocatable cargo/vehicle inspection system were successfully developed. In December 1999, the fixed large container/vehicle inspection system was put into operation at the customs office in Donggang, Tianjin. In May 2001, Nuctech signed the contract of the "H986 Project's Second-phase Mobile Cargo/Vehicle Inspection System." All of these actions have laid a foundation for the marketization and the industrialization of the cargo/vehicle inspection system. Besides, while Nuctech was opening the international market, the Chinese government also supported and took part in the action, which provided great support for the development of Nuctech.

3. Summary

In accordance with Tsinghua University's motto of "Self-discipline and Social Commitment" and Tongfang's motto of "Responsibility, Exploration, and Surpassing," Nuctech insists on being people-oriented, making the S&T development and serving the society its task, making common progress its pursuit, "Innovation, Striving, and Integrity" its philosophy, and becoming "No. 1 at home and in the world" its goal. Nuctech focuses on national and social demand, provides high-tech products, and strives to serve the national, social, economical, environmental, life, and product security. Finally, Nuctech moves along the road of collaborative cooperation of the university and the enterprise which is suitable for its development.

Enterprise Case 9

Chery—Creating the independent brand of the Chinese auto industry

1. Introduction

Chery Automobile Co., Ltd. (Chery) was founded in 1997 in Anhui Province by five local state-owned investment companies. By 2016, 5 million Chery cars have been sold. Currently, Chery has an annual output of 900,000 vehicles, 900,000 engines, and 800,000 sets of gearboxes. The company owns four sub-brands of Chery, Riich, Rely, and Karry, and its products cover passenger cars, commercial vehicles, and mini-cars. Chery now has 12 major series and 26 models on the market, and achieves a virtuous cycle of "R&D, production, and reserve," which ensures the continuous upgrading of technology and products. Chery makes "safety, energy conservation, environmental protection" the goal of its product development, and has successively passed the international quality system certification of ISO9001 and the ISO/

TS16949 of TÜV.

In 2010, Chery achieved global sales of 682,058 cars, a yearly increase of 36.3%. It became the Chinese own-brand sales champion for 10 consecutive years. Over many years, Chery has constantly promoted an internationalization strategy, and has entered the brand development stage of the "technology leader of Chinese auto brands with an international vision and trustworthiness." Its products are popular with consumers all over the world. In the same year, Chery exported 92,000 cars, a yearly increase of 93.2%, and the accumulative total exports were more than 500,000 cars. It became the largest passenger car exporter for eight consecutive years.

In its independent innovation and international development, Chery attaches great importance to the building of a brand. Through continuous improvement of "technology, quality, and service," Chery improves its brand image and production value. In 2006, Chery was recognized as China's Famous Trademark and included in the list of "China's Top 100 Most Valuable Trademarks"; in 2010, it was named "The Most Admired Chinese Company" by *Fortune* for the fifth consecutive time, and was nominated in the list of the "TOP 10 Most Competitive Chinese Companies in the World" released by Roland Berger Company; in 2011, it was listed in Hurun Brands List of the 100 most valuable Chinese brands for the first time.

2. Chery's independent innovation

2.1 Hardship at the start-up

At the beginning, Chery only had a team of eight people, and the first production line was bought from the British Ford Motor Company for 25 million dollars. In 1997, the team members increased to 50, and through continuous R&D, in May 1999, the Chery ignition engine was officially produced and was a success. In December, the first vehicle was produced. In May 2000, Sichuan Jieshun Co., Ltd. became the first dealer that purchased the first 100 Chery cars. Since then, Chery cars have entered into the market formally. In the same year, Chery was ordered to stop producing by a national authority for its not being listed in the National Car Directory. Finally, Chery had to negotiate with SAIC, and then give 20% of its registered capital to SAIC for free in order to get a "pass." In 2001, Chery was listed in the Vehicle Production Management Directory, and thus its production gradually got onto the right track. At the beginning of its entrepreneurship, Chery carried out imitative innovation, while the capital, technology, and talents were relatively scarce.

2.2 Steady development (2001–2006)

In 2001, Chery started its second stage of entrepreneurship. The representative event was that Chery commissioned the Austrian company AVL to design 18 kinds of engines which could meet the Euro Ⅳ emission standard from 0.8 to 4.2 liters during 2002–2003, and these engines were gradually adopted in Chery models in 2004. Meanwhile, Chery and Taiwan Fuzhen Company established a joint venture to gradually master the mould manufacturing technology. Chery's engine and mould projects were its preliminary preparation for rapid growth. In 2001, Chery cars were launched officially on the market, and 28,000 had been sold as a single brand in that year. In 2002, Chery's annual production and sales exceeded 50,000, and thus it successfully became top eight in the domestic car industry. In the same year, Chery invested 700 million yuan in introducing the most advanced painting line from the well-known German professional manufacturer Durr. In 2003, Chery launched three new models: QQ, Eastar, and Cowin, and sold 90,367 cars throughout the year. Since then, Chery had completed its product line layout successfully and gradually entered a new stage of comprehensive development. In 2004, Chery sold nearly 90,000 cars. In 2005, Chery increased the sales to 180,000 by reducing prices, and as a result, it withstood the first challenge of the market. In the same year, Chery achieved a great breakthrough in the development and manufacture of core components, which was also a significant breakthrough of the country, and produced the first high-performance engine. In 2006, Chery owned 5 series of models, from microcars to SUVs, so that Chery had a more complete product structure. In the same year, with the birth of the 500,000th car, the first batch of Chery engines were exported to the US, a precedent of Chinese own-brand engines exported to developed countries in large quantities. Chery had grown to be the pillar of Chinese own-brand enterprises through continuous innovation, and become one of the mainstream car companies in China. In the stage of steady development, Chery combined collaborative innovation with independent R&D innovation.

2.3 Strategic transition development: Since 2007

In August 2007, after 1 million cars had come off the production line, Chery began to enter the new stage of forging an independent international brand, followed by Chery's "strategic transformation." Whether its strategic transformation was a success or failure was inconclusive, but with the change of the enterprise's development strategy, Chery achieved a good start for its "perfect transition."

2.3.1 From focusing on "faster and better" to "better and faster"

Chery spent 93 months on increasing the cumulative production from one to 1 million, which is a record in the auto industry in China and even in the world; Chery's speed has become synonymous with the new force of the Chinese automobile. With a rapid development, the brand "Chery" has gradually been accepted by consumers. When entering the new development stage and facing further pressure to develop its domestic and international market, Chery's brand image needs to be further improved. The key to enhancing the brand image further lies in the quality improvement; it was thus urgent for Chery to optimize its systematical ability of R&D, supply chain, manufacturing, and marketing, so as to achieve the transformation from its original focus on "speed" to "quality and speed" by improving the quality and enhancing its brand image.

Chery proves its determination to improve the quality of the automobile by action. In September 2007, Chery declared that it had entered the phase of strategic adjustment which included the price, quality, service, and brand. When the major manufacturers reduced their prices in order to sell more cars, Chery implemented the "price stability" strategy. While actively slowing down the speed of competition, Chery improved the quality of its products through internal production and quality management, improving the selection criteria of its supply chain, and other measures. These measures may affect sales in the short term, but quality improvement and brand-building are the keys to independent brand development in the long run. In 2007, Chery's market performance directly reflected the result of its strategic transformation. According to statistics, Chery's cumulative sales were 381,000 with the annual task completion rate of 97% in 2007. Compared with 2006, there was an increase of 24.8% While slowing down speed and improving quality, Chery had maintained a steady developmental trend. In 2008, Chery's developmental mode changed from extensive to intensive growth by improving its quality, brand, and service, pursuing a long-term and healthy development. Through continuous efforts at improving the quality of its products, adjusting the direction of its development and improving the company's competitiveness, Chery obtained sustainable and healthy development.

2.3.2 From "keeping a foothold at home" to "embracing the world"

Facing the global market competition, internationalized development is the inevitable choice for auto enterprises after the strategy of "keeping a foothold at home." Globalization is the only way that Chery could further expand the development of its own brand and even create a well-known international brand. In 2007, Chery began to conduct its strategic transformation from "keeping a foothold at home" to "embracing the world." Since the first batch of 100 Chery cars were exported to Syria in 2002, Chery cars have been sold in more than 80 countries and regions. Chery sales on the overseas market doubled again in 2007, a sharp increase of 132%, and its annual total exports exceeded 119,800. The proportion of Chery's overseas sales in its total sales increased from 16.7% in 2006 to 31.4% in 2007; the overseas market had become Chery's significant growth sector. Up to 2016, Chery's accumulative car exports were more than 1.1 million, and Chery became the largest Chinese passenger car exporter. Since 2007, Chery's export strategy had changed from exporting products and technology to setting up overseas factories. Chery has 14 CKD (completely knocked down) factories in the world; CKD and local assembly have become the mainstream of Chery's product exports, and the proportion has reached 80% of the total export. While expanding the market further, Chery has gradually mastered the initiative of overseas market development.

Internationalization is a must for the survival and development of Chinese auto enterprises. Every enterprise has a different internationalization trajectory; Chery was the first one to explore the trajectory of international capital and technical cooperation in Chinese auto enterprises, and expanded its global market by relying on the influence of technology. Chery stepped onto the global market completely via technical and capital cooperation with first-class enterprises. This is a new exploration for Chinese auto enterprises, and fully reflects the foresight of Chery's developmental strategy and the international influence of Chery's technology. Since 2007, besides the cooperation with some world-famous enterprises like Delphi and ArvinMeritor in the fields of R&D and parts and components, Chery also carried out cooperation on capital and technology with Quantum, Chrysler, Fiat, and other international giants. Chery's international cooperation is throughout the entire industrial chain, through which Chery has enhanced its independent innovation capability, broken technical barriers, and caught up with and surpassed the advanced world level to occupy both the domestic and overseas markets. Based on its domestic market, embracing the world has been Chery's strategic goal for quite a long time.

3. Chery's trajectory towards independent innovation: From introduction, integration, original innovation, then to full innovation

Since Chery began to imitatively develop Fulwin by integrating the Ford CVH engine, it has integratedly developed Eastar (equipped with a Mitsubishi engine), QQ (by integrating a Dong'an engine), Cowin (equipped with a Chrysler engine), and Tiggo (by integrating a Mitsubishi engine) successively, and has consequently mastered product architecture technology. Later, Chery developed 18 engine models of ACTECO Series through cooperation with the Austrian company AVL in order to improve the existing models, and thus mastered engine technology. On the basis of key technology, Chery inherits innovation, and finally creates an independent brand in the Chinese auto industry.

3.1 From progressively catching-up of the core element to the building of an independent brand

Chery's independent innovation progressively went through the process of catching-up of the core element to the building of its independent brand. At the beginning of its entrepreneurship, Chery carried out imitative innovation and cooperated with professional companies to improve its technological innovation ability. The prototype of the first Chery car is the Spanish Toledo, which is in the same grade with Santana, but the price is almost a third lower. To break the technology monopoly of multinational companies in key components, Chery took the innovation system of spare parts as a breakthrough. On the basis of introducing technical talents with international experience, Chery invested, held shares, and participated in a number of enterprises dealing in key parts. So the industrial chain of independent design, production, and marketing of key components was initially established. Its independently-developed intelligent sunroof and many other technologies filled in the gaps in the country, and some had reached an international level, which greatly reduced the cost of the vehicle and laid an important foundation for the development of own-brand cars. In 2005, Chery's first high-performance engine came off the assembly line, marking Chery's great breakthrough in the development and manufacture of core components; Chery had gradually realized the progressive catching-up of the core elements.

Since 2006, Chery has gradually established its own brands, including the new Cowin, Tiggo, QQ, Eastar, etc. At the end of 2007, Chery had become one of passenger car enterprises owning the most products and the most complete product series. With the development of the enterprise, Chery has completely mastered automobile core technology, built its own brands, and gradually progressed from catching-up of the core element to establishing its independent brand.

3.2 Improving the design, upgrading the product, and perfecting the independent brand according to the market

Chery is dedicated to improving product design, breaking through the low-end blockade, upgrading its products, and perfecting its independent brand according to the market. As for improving the production design, a typical representative is the overall security design of the Chery Fulwin. The model in 2005, at the price of 59,800 yuan, was equipped with dual front airbags and ABS (antilock brake system), which were key elements of active and passive safety equipment. This was absolutely second-to-none in the industry. In 2003, Chery's ABS was the overall package of ABS plus EBD (electronic brake force distribution system) plus ESBS (extended stability brake system), which ensured the tracking of the braking system and the stability of high speed. It can be said that Chery created a new safety standard for cars in the price range of 60,000 yuan. The comprehensive safety design of the Chery Fulwin promoted product quality, and further improved its independent brand. Regarding product updating, there were some typical examples, such as the proposal of a digital QQ and the launch of Eastar 2005. The digital QQ came from the the upgrading of QQ 3, and its dashboard changed to an LCD digital dashboard from the original mechanical one. With regards to the configuration, the digital

QQ was equipped with a parking sensor and a remote CD. Concerning the engine, the sound insulation of the engine was strengthened. In addition, the whole car's electric circuit was equipped with a fuse box, which is convenient for future maintenance. The Chery Eastar 2005 was officially launched with 60 technological improvements on the basis of the old model. In short, Chery never gives up the brand with good sales performance, and keeps improving its quality through putting new products onto the market.

3.3 The expansion of an independent brand—From "a world-class car on the Chinese market" to "a Chinese car on the world market"

In 2008, aiming at forging an "independent international brand" and relying on its technological advantages, Chery cooperated with world-class enterprises on developing the global market and achieving the expansion from "a world-class car on the Chinese market" to "a Chinese car on the world market." It is believed that auto exports are the indicators that prove the level of a country's auto industry. Only by bulk exports can an auto enterprise step into the high-level competition in the global auto industry. Chery has proven its ambition and determination to step into the global auto market since its foundation. Chery passed the on-site test carried out by the German company TÜV in August 2008, which made it the first vehicle manufacturing enterprise to pass the most stringent and advanced automotive production quality control system in the global range—ISO/TS16949. In October 2002, the SAP/ERP project implemented by IBM officially came into operation. The construction and use of SCM and the use of the first phase of the CRM project allowed Chery to achieve synchronization with the international standard in enterprise management and working procedure. Then in 2007, Chery set up a joint venture with Quantum and signed cooperation agreements with Chrysler and Fiat. As Chery developed, the promotion of its independent innovation ability aroused the interest of the world's largest auto companies, such as Chrysler and Fiat, to establish long-term cooperation with Chery. The cooperation covers technical design, model development, and plant construction, as well as the procurement system. Chery has gradually gained equal discourse power in international cooperation and has started a new mode of global auto cooperation. And as a result of its development in technology and product quality, its products are finally exported all over the world. Until 2010, Chery had been the largest export enterprise of passenger cars for eight consecutive years, and gradually realized its aim to sell "Chinese cars on the world market."

3.4 Innovation of the marketing mode

The key to increasing product sales and realizing the value of independent innovation is innovation in the marketing mode. In order to increase its production, innovation of the marketing mode is carried out throughout Chery's entire process of development.

In March 2005, Chery experienced a great reform in its sales method. Chery separated its products sold by hundreds of agents to sell in different networks. This sales mode keeps itself in good order on the market and maintains the stability of the price, from which agents can also gain benefit. Chery cultivates a great number of agents with the help of its new marketing mode, which has paved the way towards its rapid development. In 2007, the new marketing mode went into effect, and three Chery auto towns (Iseway of Xi'an, Lianhai of Shanghai, and Chengxingda of Beijing) were completed, which proved that Chery had made a further more profound step into intensifying its channel. Since then, Chery has stood more firmly at the command of the market in northern China. Later, Chery put forward new innovative marketing modes, guided by "Super Drill Plan," "Wild Goose Team Structure," and "One-one Project"; Chery also put forward the plan called "Across China" in order to establish a cluster of super 4S stores. The implementation of this plan demonstrated that Chery had come into a new period of the marketing mode of Chinese cars, and functional escalation and intensification became a new trend in the whole auto industry at that time. The "Across China" plan is a new marketing mode, which has created a closer connection between Chery and its agents; it helps Chery adapt to the rapid development in the future and calmly face the fierce competition on the global car market. As a result, Chery will be able to create a distinctive brand in the long run.

3.5 The protection of independent IPR that accompanies the independent brand

Protecting independent IPR is an important part of independent innovation and the key to ensuring enterprises, individuals, and other innovation subjects continually bring forth new ideas. Chery spares no effort in the protection of independent IPR. First, it actively applies for patents. Chery was the top among local auto enterprises due to its 12,882 patent applications and 8,129 authorized patents by the end of 2015. Second, Chery has built into its enterprise an information security system in case of leaks. On the one hand, it has put up an intranet safety management and digital IPR protection system; on the other hand, it has strengthened its humanistic management. That is, the security management of core data, of recorders and portable storage devices, of application programs, and of internet applications. These managements involve every aspect of the daily work of ordinary staff and play a vitally important part in preventing the leak of internal data. Due to various kinds of methods and measures to protect independent IPR, Chery built a wall of safety and security.

4. The features and implications of Chery's independent innovation

After years of rapid development, Chery has become the largest enterprise to produce own-brand cars with the greatest comprehensive strength. In addition, Chery has become the first enterprise to export vehicle manufacturing technology and CKD parts. What makes Chery go so far in car marketing? And what are the features of its innovation?

4.1 The progressive catching-up based on core technology that makes Chery turn from imitation, improvement to independent innovation

Chery, after surviving in the market, realized that independent innovation was the foundation of an enterprise. Only by mastering core technology and cultivating core competitiveness can an enterprise develop its own brand. With the basis of a stable R&D investment, Chery experienced the process from imitation, improvement, to building an independent brand. From the Fulwin model based on the Spanish prototype Toledo to the first independently produced and high performance engine, the digital QQ, Tiggo, etc., Chery achieved its technical development and leaped in the process of acquiring core technology, which protected Chery from being dominated by others, and promoted its sustainable and healthy development.

4.2 Mastering the core technology of key components and forming a sound independent supporting system

In order to break the monopoly of multinational companies and reduce costs, Chery has taken an active part in developing an independent innovation system of key components and material. In addition, Chery has introduced the outfits and means of manufacturing and detection at the advanced international level. Combined with its own adoption and innovation, Chery has created the best conditions for ensuring good quality products and carrying out independent innovation in the most possible way.

4.3 The global human resources strategy and cooperative development that enhance Chery's independent innovation capability

During its development, Chery has recruited talented people from all parts of the world and attracted a wide range of well-known technical experts in the auto field domestically and internationally. Thus, Chery has carefully created a "multinational force" under the principle of gathering the world's best human capital. Its rapid advancement of independent innovation has attracted a large number of technicians and experienced management personnel from the world's leading auto companies, parts firms, and major domestic auto companies; this helped young Chery form its own mature technology and management team. Such a "multinational force" and other engineering and technical personnel made up an R&D team of 6,000 people. This team undertakes Chery's major R&D task, which involves the vehicle body, the power assembly, the chassis, and the electrical appliances. This team continuously strengthens Chery's independent innovation capability.

Chery chooses to carry out R&D with world-class technical companies and institutes, but not to produce automobiles with multinational joint ventures. This is a unique innovation mode of Chery. In the R&D team, all products are defined by Chery, and the technical routine is decided by both sides. As a result, Chery can, on the one hand, graft technologies and train personnel, and on the other hand, obtain complete independent IPR and then form its own "knowledge accumulation." Through independent R&D dominated by international cooperation, Chery preliminarily mastered developmental technology and its process and then got a batch of achievements at an advanced international level in key technologies such as the engine, the model, the chassis, and whole vehicle development.

4.4 Chery's unique development strategy of independent innovation

Since its foundation, Chery has insisted on developing its autos and key components with independent IPR. In the process of independent innovation, all the innovation development strategies promote Chery's growth.

(1) Going out—Joint development and completely mastering core technology. The most representative issue is the joint development of Chery and AVL, an Austrian engine development company. The product developers from Chery play an important part in the overall process. Besides, there is a training plan for the personnel who are engaged in product design, testing, assembly, and calibration, so that Chery can fully master the core technology.

(2) Inviting others—Joint development and cultivating its own R&D team. With the aim of mastering the core technology of the whole vehicle and engine, Chery invites top talents from foreign design companies to take charge of program management and design together with young engineers from Chery. Besides, Chery sends a large number of engineers abroad to participate in the overall developmental process. Therefore, Chery cultivates its own R&D team.

(3) Two sides and one center—The innovation of the concept of development. Chery divides vehicle development into many parts. The conceptual design will be finalized by design companies, while the engineering design will be completed by Chery itself, and the final examination and check will again be carried out by design companies. In order to reduce costs, Chery will negotiate with only one design company concerning one or more models. According to the advantages and disadvantages of the different companies, Chery will sign different contracts with different design companies to optimize the design in terms of different stages. For example, Italian design companies are skilled in modeling, so Chery will invite them to design the appearance. Then Chery turns to British companies to test the products. In this way, Chery can achieve the optimal design.

(4) Gradual accumulation to form the R&D system with its own characteristics. i) Chery has built a new mode with the Automotive Engineering Research Institute as the core and carried out development with all kinds of resources. For product R&D, Chery has set up a coordinated product R&D system centered around the Automotive Engineering Research Institute. This system has four branch institutes (two domestic in Shanghai and Beijing, and two international in Tokyo and Turin); and its backbone is a group of holding design companies. These institutes design collaboratively along with component companies and suppliers. Meanwhile, Chery also cooperates with research institutes of domestic universities and colleges. So far it has made a complete development system from the design to the testing of the whole vehicle, power assembly, and key components. Moreover, Chery has built its own standards system and database. ii) By integrating international resources, Chery establishes its R&D institutes in the areas with the most advanced S&T. It is important to Chery's method of integrating resources. For example, Italy is the shrine of automotive modeling design, so Chery has invested in and set up companies that deal with whole-vehicle design and automotive engineering design as well as design consulting companies in Italy. Senior modeling and structural engineering design engineers along with project designing management experts are employed to bring the most advanced design concepts to the designing of Chery's new models. In order to improve the level of manufacturing process, especially to meet the demands of technological level promotion on the production site, Chery has set up R&D and consulting companies in Japan to integrate the resources of the development and production management of Japanese vehicles. At the same time, Chery seeks Japanese automobile professionals and selects some to improve its technological level of the production site. By integrating international resources, Chery improves its capability for independent R&D, and the development cost is also reduced.

5. Summary

Chery sets a new benchmark for the development of Chinese auto enterprises, and is a prelude to the next boom of independent auto brands. The key to its success lies in its unique innovation concept and system. Chery has shaped the all-embracing innovation system to take advantages of the resources at home and abroad by technology import and cooperation. Such a system helps Chery seize the initiative in international cooperation and master management in core technology. Of course, its passion to be top-class in the world also counts.

Industry Case

The bottleneck and trajectory of the localization of the Chinese equipment manufacturing industry

The localization of the major equipment manufacturing industry is related to Chinese industrial security. Former General Secretary Hu Jintao emphasized that improving independent innovation capability was an important national policy of our country in the report of the 17th CPC National Congress in 2012. How can the ability of independent innovation be improved? What is the purpose of independent innovation? What is the relationship between localization and independent innovation? What kind of consequences are there, if we only have independent innovation without the localization of technology? These questions pose a severe problem for the capacity of independent R&D and for the systematic integration of the equipment manufacturing industry and the strategic manufacturing sector. The equipment manufacturing industry is a major industry for the national economy and for the livelihood of people, so its localization is of great significance. However, the low-level localization of the industry is always a bottleneck for Chinese modernization. How to break the bottleneck in order to improve the level of localization is a key problem.

Based on the research regarding iron & steel machinery, electromechanical machinery, petrochemical machinery, paper-making, sewage treatment, flue gas desulfurization (FGD) in the coal-fired power plants, nuclear power lifting, thermal power units, air cooling systems, the textile and construction industries as well as the high-speed railway, we will try to investigate the reasons and put forward solutions to these problems.

1. The equipment manufacturing industry and its characteristics

Major technical equipment refers to the knowledge, technology, and capital intensive products. The major technical equipment manufacturing industry is a historical category. With the evolution of history, its specific content has changed accordingly. Why to develop the major technical equipment manufacturing industry is a strategic national issue. And how to develop it is a route strategy, depending on the country's strategic objective and the international situation; localization is one of the focal points. The major technical equipment industry has the following characteristics: High technology, complicated production, strict process requirements, high level of completeness, complex coordination relationship, and high correlation. The equipment manufacturing industry is the reflection of the national comprehensive manufacturing capacity. And the ability to build up a major technical equipment manufacturing industry is an important standard to measure a country's industrialization and comprehensive national strength; it is also an important means and industrial carrier for high technology to transform into productivity. In the process of industrialization, the equipment manufacturing industry has the laudatory name of the "equipment department of the national economy." The major technical equipment manufacturing industry is the core part of the equipment manufacturing industry and the "mother ship" with the highest level of technology, the most extensive and complex collaboration, and the most difficult integration system.

2. The situation and problems of the localization of the Chinese equipment manufacturing industry

2.1 The development of the Chinese equipment manufacturing industry

As early as 1986, the State Council organized a seminar around the theme of "buying or developing modernization." The former Minister of Machinery Ministry, He Guangyuan, clearly stated that modernization cannot be bought, and that we must develop independent brands, revitalize the national industry, and finally achieve modernization. 20 years later, the leaders and experts who were engaged in the revitalization work of the equipment manufacturing industry sat together once again in June, 2006. What they had discussed was still the topic—to realize the localization of the major technical equipment as soon as possible and give the Chinese equipment manufacturing industry an opportunity for development.

The level of localization of some equipment manufacturing industries has been greatly improved in recent years, such as the high-speed railway, aircraft manufacturing, aerospace equipment, etc. But there is an indisputable fact that the localization rate of most of the equipment manufacturing industries is extremely low and the import rate stays high. The import value of steel rolling equipment was as high as 1 billion dollars in 2004, accounting for 67% of the metallurgical equipment imports. However, the domestic equipment manufacturing industry made up more than two thirds of the weight of metallurgical equipment, but the value was less than one third that of the equipment. Although some equipment does not need the introduction of a complete set of equipment, some single pieces of equipment continue to be introduced, such as the bell-less top (BLT) and coke dry quenching (CDQ) of iron in the metallurgical industry. Although the Chinese technical level is not poorer than that of other countries, the introduction of these equipments never stops, so that the investment in introduction remains high in the metallurgical industry in China in recent years; sometimes the rate is up to 45% of the total investment in the industry. Among the introduced products valuing more than $20 billion in the electrical engineering each year, the largest importation is the T&D facility which spends more than $10 billion, mainly on four types of products: Large transformers, high, medium, and low voltage switches, wire and cable products, and control, protection, and automation products, including ballast devices. The highest import amount of products among the four categories is high voltage switches. From January to April, 2008, the Chinese total importation value of electromechanical products was $129.56 billion with a year-on-year growth of 28.5%. Among them, the importation value of excavators was $362.44 million with a year-on-year growth of 78%; wires and cables, $990.38 million, 22%; electron components, $40.37575 billion, 34%.

Meanwhile, two-thirds of the equipment in Chinese fixed assets investment depends on imports; the equipment market for optical fibers, integrated circuit chips, petrochemical, vehicle, numerical control, textiles, and offset printing is dominated by imported products. The foreign trade deficit of the equipment manufacturing industry in China is as high as tens of billions of dollars every year. These facts put the revitalization of the equipment manufacturing industry into an awkward position, and even threaten the safety of the Chinese equipment industry. Now, a lot of key equipment in monopolized industries depends on imports, which worries us.

2.2 The essential problems of the low level of localization of the Chinese equipment manufacturing industry

First, the first (set of) equipment gets few chance to be used. The characteristics of the equipment manufacturing industry, such as huge investment, long construction, and high requirement for accumulated experience, make it essential that the first (set of) equipment should be recognized, but at present the bad market performance of the first (set of) equipment in the Chinese equipment manufacturing industry has become the critical defect that keeps the local equipment enterprises from participating in the market economy. If the first equipment has unsatisfied performance, the manufacturing enterprise not only loses its market opportunity in the future, but also gets a serious blow to its confidence, followed by being "specialized accompanying-bidding participant," and serving the international companies. Its low price not only gives a bad impression, but also exacerbates the low-end melee of the market without competitive advantage. Or else it can change its profession and become dependent on international companies, and this further weakens the power of the industry.

Second, domestic users do not trust homemade equipment. Customs' import and export statistics show that in recent years, with the steady improvement of the Chinese equipment manufacturing level, the quantity of imports and the amount of major equipment have not decreased; instead, they have increased year by year. Some people believe that Chinese equipment products cannot meet the needs of the rapid economic development, but in fact, this is not true. To a great extent, many domestic equipment products can replace imported products, and some devices even have exceeded the advanced international level, but due to the prejudice, imported equipment wins out over domestic equipment in the domestic procurement biddings.

Third, multiple and repetitive introduction never stops. The long-standing lack of the macroeconomic regulation and control of the introduction of complete sets of equipment results in the fact that multiple and repetitive introduction cannot be stopped, and the equipment manufacturing industry has fallen into the weird cycle of introduction–relapse–re-introduction. There is an information asymmetry between monopoly sectors and local enterprises; the former block information and keep secrets regarding the concrete parameters of imported products, so that the latter are never able to catch up with their brand. There is no chance for them to breathe before they fall. They cannot get market application opportunities or accumulate experience, and the less experience they have, the lower their level is; the poor level leaves them no users, and all of this causes a vicious cycle. Of course, there is no denying the fact that the Chinese domestic equipment industry has its own problems: Incomplete variety, unreasonable structure, low technical level, unstable quality of the products. Because of the high cost of basic construction, domestic enterprises try to save the initial investment and adopt the procurement mode of foreign companies—to supply less and design more; as a result, the conservative design leads to great waste.

Fourth, equipment advancement is pursued, but the matching is ignored. The major equipment manufacturing industry has a high requirement of matching and environmental adaptability. Some must be attached to the original "mother ship." It is not true that the most advanced equipment achieves the best effect; it must match, otherwise it will cause an overflow of the function, which is a waste. However, some enterprises blindly pursue the latest and the best, which causes no effect except wasting money. At the same time, it exceeds domestic engineers' adoption and implementation ability.

3. The analysis of the cause of low-level localization of the Chinese equipment manufacturing industry

3.1 The users' monopoly in the equipment manufacturing industry is the main reason that localized equipment is abandoned

From the perspective of economics, the market and products are as close as lips and teeth. With regards to the confusion over the development of the domestic equipment manufacturing industry, many experts think that "some domestic users, especially big users have a certain monopoly (large state-owned monopoly operating institutions for example) and do not use domestic equipment," which causes great resistance in the development of the industry. Compared with some users with a higher degree of monopoly, the equipment manufacturing industry is already a market-oriented one. Since the early 1980s, the price of mechanical products has been completely determined by the market, and almost no product is priced by the government. The past pattern was the 333 mode (state-owned, joint venture, and private capital, each with one-third). However, after the reform of state-owned enterprises, the equipment manufacturing industry has made a big step forward. Many state-owned enterprises are not traditional ones in a sense; some even launched IPO at home or abroad (for example, Harbin Electric is listed in HKEX, and Harbin Air Conditioning in SSE). Great changes have taken place in the ownership structure of the industry, but its users have not radically changed, especially some monopoly users in the true sense, such as petrochemical, chemical, oil, T&D, nuclear power, railway, mining, and nuclear power sectors, etc., but there is a severe problem of information asymmetry. First of all, there are large power subjects with an impalpable character that exist between the plan and the market, between the government and enterprises. Second, the market of the Chinese strategic resources and infrastructure network industry presents a special kind of market structure, namely "open at both ends and monopolistic in the middle." Third, this kind of monopoly operation or management organization institutions are the products and promoters of a gradual reform of the economic system, which is closely integrated with the national policy system and even the government regime. Finally, it is difficult to fundamentally eliminate their interference in the market mechanism. Under this management system, the following issues might arise.

Fear of accountability: The safety of production and operation is the first criterion used to assess leadership. When there are accidents in the operation, domestic equipment is to blame, while foreign products have no responsibility. Multinational companies are not controlled by the administrative system, so the government has no right to blame them.

Fear of failure: Regional segmentation among departments objectively results in not forming a mechanism which can shoulder the risk of independent innovation.

Fear of one's own losses: Monopoly operations do not have good financial commitments. Their budget is not only the constraint of the cost and profit, but the balance mechanism of the monopolistic profits and the off-the-book benefits.

In fact, a more significant reason is that the money is from the state budget, making equipment quality not important; worse, the government does not allocate the money saved by purchasing domestic equipment, but when there is a problem with the purchased foreign equipment, the government will handle it with money. The fundamental reason is the privilege of the industrial monopoly, which always provides many monopolists with a countermeasure to the government policies. The Chinese government has announced a lot of policies regarding localization and independence, such as the ban on importation of complete machines; however, the equipment procurement sector is always able to come up with various countermeasures. For example, they will require the local equipment factory to import most parts from specific countries as a precondition. On the one hand, they can get a tax rebate on the importation of key components and be exonerated from all kinds of tariffs for the importation of the entire machine. On the other hand, they can enjoy all the benefits of buying foreign brands and become a model in the purchasing of domestic equipment.

At the same time, through the structure of the bidding process, they can purchase foreign products. They directly list the products of some models and performance. For example, a power plant will directly require foreign GEA and SPX equipment procurement in the bidding on desulfurization equipment; GEA and SPX have almost monopolized the domestic market of air-cooled condensers in large-scale power stations. So they exclude local companies by the specific requirements and higher bar. The Chinese users' dependence on foreign equipment directly leads to the monopoly of foreign enterprises. Disguised importation has become an important way for some users to repel domestic equipment.

In addition, "the user is the king" in the major equipment industry chain; the largest profit belongs to the user. The local equipment industry spends everything on giving up its biggest profit space in order to motivate the users to purchase their equipment and even help them set up projects, but they still do not have any interest in local equipment. In this sense, the key issue is not the quality and the technical level of local manufacturing equipment, but the fact that only when the monopoly sectors rely on the technology and products of multinational companies can they maintain the interests of the monopolistic industry. No users, no market pull. How can we talk about the updating of major domestic equipment applications and technology? How can we talk about the revitalization of the whole equipment manufacturing industry? No usage is a big problem in the revitalization of the equipment manufacturing industry.

3.2 Localization is difficult to be achieved because the segment of the industrial chain results in no intersection between demand and manufacturers

The analysis above shows that the industrial chain of the Chinese equipment manufacturing industry is segmented. The upstream enterprises, namely the equipment manufacturers, tend towards marketization, but the equipment users are mostly state-owned monopolists. Because of its monopoly and political nature, it is difficult for the upstream enterprises to extend to the downstream, and the downstream enterprises disdain the upstream line. If there is no possibility to reconcile, it is difficult for both sides to form a community of interests. Consequently, the complete industrial chain is broken. And the following problem will appear. The side that needs the equipment starts from its own interests in the introduction–re-introduction cycle; while the equipment manufacturing side follows the cycle of introduction–adoption and implementation–re-innovation–independent manufacturing–the first equipment not being used–re-introduction–re-innovation–the equipment still not being used. In such a technical trajectory, the upstream and downstream do not intersect. What is worse, equipment users keep secrets about the volume, places, parameters, and indicators of imports for their commercial interests; however, the equipment manufacturing side cannot introduce a whole prototype machine by spending a huge sum of money just for adoption and implementation. According to the survey, some of the Chinese iron and steel production enterprises even reject domestic process control technology and new products, which greatly delays the localization of metallurgical equipment. The information asymmetry between the upstream and downstream enterprises makes the equipment localization unable to keep pace with the equipment users' introduction and thus falls into a vicious cycle.

Some leaders of the equipment users even claim that localization is the responsibility of the production departments themselves. The most important reason for this is that there is no integrated national planning of the cross-sectoral and cross-regional industry. As to improving local advantages of the industrial strategy, there is little cooperation among the provinces. The arrangement and combination of the industrial planning of many machinery sectors lack key points and fine order; there is almost no unified strategic target system and no institution like the Japanese consortium to reconcile and command. The monopoly sectors' bulk purchase of foreign technology or equipment has a short cycle and obtains a quick effect in the short run. However, it will bring about tremendous obstacles to catch up with the foreign equipment manufacturing sectors in the long run. Monopoly users already plan to buy the next generation of foreign equipment, when the local companies still have not completely understood the technical performance and indicators of the imported equipment.

There are also some equipment monopoly users who are "following the fashion and pursuing the so-called efficiency." Once there are new products on the international market, under the pressure from the foreign party and also because they are subject to the interests of relevant sectors, for example, in order to receive their $200 subsidies for going abroad, some specific users will implement the so-called strategy of "attracting foreign capital" at the cost of $2 million from the state budget in some mechanisms. They are eager to purchase new products without considering whether they are suitable to their needs, whether the workers are able to control the new equipment, or whether domestic production equipment can also meet their requirements. All in all, besides ignoring the local equipments, they also complain that their quality and performance is not up to standard.

3.3 The original investment and financing system, fiscal and tax policies, and their execution rules are in contradiction with independent innovation

Taking tax preference as an example, in the 1980s, China gave a lot of "green lights" for investment attraction and capital introduction, and provided a security policy of "no tax for the first 3 years and half tax for the next 5 years" to the foreign-owned enterprises or joint ventures, as well as providing free taxation for the importation of entire machines needed by industrial development, which, however, did not include the parts. This series of preferential tax policies had continued for more than 20 years. While China encourages independent innovation and localization and gives the same national treatment to domestic and foreign enterprises, it is difficult to get rid of Westernized policy and start all over again. A super-convenient national treatment to foreign investment is still a pain for domestic enterprises, which is one of the main obstacles that impede the revitalization of the equipment manufacturing industry.

3.4 The contradiction among leadership tenure, performance evaluation, and the long cycle of technological innovation and industrial development

This problem seems to have no direct link with equipment localization, while in fact it does directly affect the process of equipment localization, especially in the major technical equipment sector, the structural adjustment, the upgrading, and the completion of major projects that require at least 10 to 15 years or even longer, which is much longer than the leadership tenure. Consequently, there is often a lack of continuity. In the past 20 or 30 years, comprehensive administrative departments, especially provincial and municipal governments, have formed the employment mechanism of assess cadres by the scale of investment and the number of state-owned restructurings. For example, some power groups make entering the top 500 an evaluation target. Thus, these enterprises can only buy ready-made advanced foreign technology and equipment for quick and better development so that there is no place for domestic equipment. The short-sight of attaching importance to quantity over quality and ignoring the integrity of the former industrial system fundamentally hinders the implementation of the independent innovation policy. In addition, the problems of the equipment manufacturing enterprise itself are also very important, including: Weak enterprise design and R&D capability, lack of money and talents; short-sighted vision; the backward state-owned enterprises' operation and management system; inflexible trajectory of equipment localization. All of these aspects hinder the localization in different degrees.

4. The analysis of the trajectory of the localization of the Chinese equipment manufacturing industry

There are lots of complex reasons why the level of the localization of the equipment manufacturing industry is low, but there is still a trajectory to follow. As long as we can find the crux, we can break it to improve the level of localization.

4.1 Analysis of the developmental trajectory of the equipment manufacturing industry

Before we choose a trajectory, we should first understand the status of the development and the possibilities.

Trajectory one, introduction–falling behind–re-introduction. When enterprises introduce the first generation of products and equipment from abroad, they will produce the first equipment after adoption, implementation, and improvement. If there is no policy support or opportunities at this moment, due to the bad popularity, reputation, and system (except for quality), the first equipment may possibly not be used in the downstream industries, which makes the first equipment stillborn. As a result, the enterprises must continue to introduce more advanced foreign equipment (the second-generation technology products). After adoption, implementation, and improvement, if their products still cannot be put into operation on the domestic market, the users will force them to further introduce the third-generation technology products, and the local equipment will never get used. Over time, the development of local equipment will fall into the situation of "introduction–falling behind–re-introduction." At present, the equipment manufacturing of many industries in the country is in such a dilemma, such as the direct air cooling equipment of power stations.

Trajectory two, progressive catching-up. The difference between this situation and the above one is that after introducing the first-generation technology, through adoption, implementation, and re-innovation, the first equipment can be trial-produced and put into operation under the efforts of all parties, and it can meet the requirements of the users. After further performance optimization, it can be quantitatively produced. Along with the accumulation of experience and absorption of advanced foreign technology, the products can catch up with the advanced international level. This trajectory of localization is called progressive catching-up, which is realistic and reliable; an example is the thermal power unit equipment in China.

Trajectory three, leaping innovation. This is a short-cut. Along with the accumulation of innovation ability, enterprises can directly introduce or purchase the most advanced technology by M&A as well as joint design. Combined with the actual situation, they can optimize the products and strive for synchronization with the advanced international level. This is called leaping or breakthrough development, and an example is the aerospace industry in China.

4.2 The realization trajectory and measures of the localization of the Chinese equipment manufacturing industry

4.2.1 The innovation of the method and process of localization of the equipment manufacturing industry

On the way towards localization, many companies complain that it is difficult to obtain foreign enterprises' advanced technology. Of course, because of the protection of IPR and the domestic enterprises' cautious attitude towards their own technology, it is difficult to obtain the technology that they need directly; but there are some flexible ways. For example, in the process of international cooperation, technology introduction, joint design, binding bid, cooperative development, and overseas training are all conducive to localization.

Based on this, the companies can optimize product performance in order to achieve independent R&D. It is the first step towards market-oriented production. Next, it is important to be recognized by domestic equipment users. The enterprises can apply for patents, obtain certification from the national authoritative department, and formulate industrial standards. In the process of bidding on the order, they should try to make the first (set of) equipment get trial production and put into operation. This is an especially crucial step. Along with the installation and debugging of the first (set of) equipment, the enterprises can finally provide the solution to the overall plan for users and realize mass production; the process of localization can be said to be "started."

4.2.2 The mode of engineering procurement construction (EPC) led by design, a better trajectory to follow in order to achieve the localization of the equipment manufacturing industry

In the mode of EPC led by design, the contractors are responsible for the design, procurement, and installation as well as the pilot-run (EPC turnkey). The design institute that is professional concerning the process and technology conducts the EPC, and can avoid the waste caused by the configuration of different levels and pay more attention to cost performance in procurement. Moreover, they attach more importance to the technological adaptability of the equipment in order to reduce the interference of non-technical factors in the bidding projects. An outstanding feature of the equipment manufacturing industry is that the equipment users adopt bidding to entrust a specialized company to complete the design of the project and procurement, and installation and debugging of the equipment. Because a specialized company is more familiar with the technical performance of equipment, and the EPC company tends to purchase equipment according to design scheme, cost, and benefits. In the process of equipment procurement, some clients will ask to know the origin and the brand of the equipment in the tender document, but most of the clients will entrust them with full powers. The equipment manufacturer needs to find the appropriate decision-maker for the equipment procurement, and convince it to use its equipment.

In addition, if some professional design companies are equipment producers, the build-operate-transfer (BOT) mode will be a shortcut to technology localization. In this way, the equipment can get used, and the operation of the service equipment can be tracked, until the turnkey, the maintenance of the equipment can be achieved from the beginning to the end, such as the project cooperation agreement for the 110 MW capacity unit of the flue gas desulfurization system signed by Ningbo Oriental Environmental Protection Equipment Co., Ltd. and Henan Sanmenxia Huineng Thermoelectricity Co., Ltd. The project adopts the "DS sulfur dioxide flue gas treatment technology" owned by Ningbo Oriental Environmental Protection Equipment, which completely owns the independent IPR, and is operated with the BOT mode. At present, the desulfurization degree of Ningbo Oriental Environmental Protection Equipment's technology can reach more than 95%, but the price of the main desulfurization equipment is only half that of the imported equipment. In the application of the technology, the materials after desulphurization are widely used as farmland fertilizer and in soil improvement projects to overcome the malpractice of using traditional limestone which will produce a large amount of gypsum and carbon dioxide. "DS sulfur dioxide flue gas treatment technology" and the implementation experience is combined with the financial operation, environmental technology and engineering. This practice widens the path and the method of cooperation in the sulfur dioxide flue gas pollution treatment in China and opens a good route towards equipment localization.

4.3 The necessity for China to set up engineering design and research institutes and other institutions to create opportunities for the equipment manufacturing industry

The repetitive and excessive introduction in the equipment manufacturing industry in China is a serious problem; we can analyze the enterprises' situation, business format, and development of the Chinese major equipment manufacturing industry by setting up engineering design and research institutes. For example, if this industry's equipment has a high localization level and the local equipment can satisfy the domestic demand, there will be no need to import.

In addition, a generic technological innovation platform in China is absent. Ignoring basic research leads to the repetitive R&D of enterprises, so engineering design and research institutes should integrate relevant research institutes to improve the capability for generic technological R&D. Also, these institutions should find out which technologies are based on the core components, which are on key materials, and which are on localization. When mastering all of the core technology, the complete machine can still not be produced. So there is no need to blindly pursue the lead level and high standard of core components. The most important thing is the breakthrough of overall efficiency and environmental protection. Engineering design and research institutes and other relevant agencies can grasp the process of localization at the macro level.

4.4 The attention paid to the localization of major equipment

The equipment manufacturing industry is related to the lifeblood and security of a country. China must pay much attention to the localization of the major equipment. At present, the Chinese government has realized this problem. In the "Several Opinions about Accelerating the Revitalization of the Equipment Manufacturing Industry" issued by the State Council, the government clearly encourages enterprises to order and use the first (set of) domestic major technical equipment. The national key project, ordering and using the first (set of) domestic major technical equipment, can be identified as a demonstration project, and can be given priority. When users buy the first (set of) domestic major technical equipment, the country can give support by providing appropriate subsidies or accelerating equipment depreciation. China will set up a major technical equipment insurance mechanism in which risks and benefits are shared with the project owners, equipment manufacturing industries, and insurance companies. The mechanism will guide equipment manufacturing enterprises and project owners to insure the use of the the first (set of) domestic major technical equipment and to support a batch of demonstration projects of key technology and equipment localization.

In April 2009, the government came up with the "Revitalization Plan on Equipment Manufacturing Industry," which was centered around the following six aspects: Setting up the special funds for the technical reconstruction of the equipment manufacturing industry; establishing scientific research projects concerning the major technical equipment; strengthening the macro-control to ensure that domestic demand for equipment products can effectively stimulate the development of the equipment manufacturing industry; prior purchasing of domestic equipment in the construction of the major projects in key areas of traffic, energy, and raw materials in the following three years; adjusting the current export tax rebate policy and canceling the interest subsidy policy regarding the importation of equipment products issued in 2007; promoting the merger and reorganization of enterprises and increasing the funding support of the equipment manufacturing industry.

Any country with developed industries will choose to use domestic equipment even if it is not "armed to the teeth." Though the domestic products might not be good enough, they will still be used. The Republic of Korea is a typical example. Almost every Korean prefers to use national products, and so do the people of the Soviet Union. This fact shows that only by "using" can the equipment production become better and better.

Index